How to Build and Modity
GM LS-Series Engines

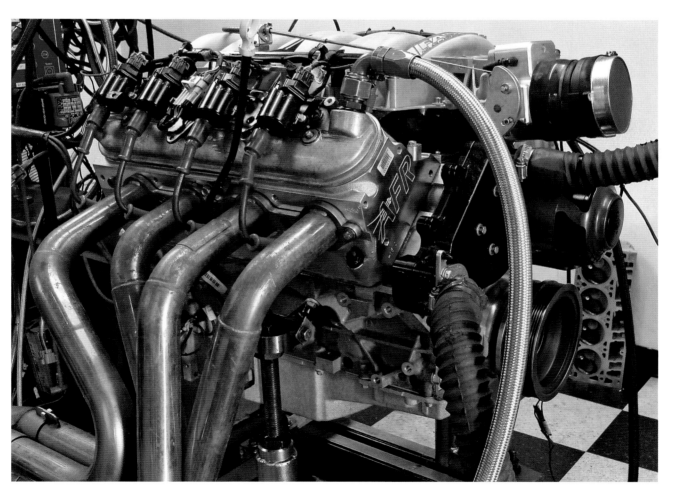

By Joseph Potak

First published in 2009 by Motorbooks, an imprint of MBI Publishing Company, 400 First Avenue North, Suite 300, Minneapolis, MN 55401 USA

Motorbooks titles are also available at discounts in bulk quantity for industrial or sales-promotional use. For details write to Special Sales Manager at MBI Publishing Company, 400 First Avenue North, Suite 300, Minneapolis, MN 55401 USA.

To find out more about our books, join us online at www.motorbooks.com.

ISBN-13: 978-0-7603-3543-7

Editor: Chris Endres
Designer: Heather Parlato

Printed in China

On the cover: GM Performance Parts LSX block's cast-iron construction will stand up to nearly any amount of power.

Inset: The Corvette Z06 is equipped with the 427-cubic-inch LS7.

On the title page: Dyno-testing your engine is the best way to ensure it performs to your expectations prior to being installed in your car.

About the author
Joseph Potak is a professional LS-series engine specialist and technician. He works for Texas Speed and Performance, catering to the performance needs of owners of LSX-equipped vehicles. Joseph is an Automotive Service Excellence (ASE) certified Master Technician and has also completed GM Factory Training pertaining to the LSX-Gen III engine family. His work has appeared in numerous magazines and can also be spotted at the drag strip, powering various LSX setups he has helped build throughout the years. In his free time, Joseph volunteers as a forum director within the LS1Tech.com community. Ironically, and despite his "LSX-only" mindset, he recently built and races a street-legal nine-second turbo Mustang.

Library of Congress Cataloging-in-Publication Data
Potak, Joseph, 1975-
 How to build and modify GM LS-Series engines / Joseph Potak.
  p. cm.
 ISBN 978-0-7603-3543-7 (sb)
 1. General Motors automobiles--Motors--Maintenance and repair. 2. General Motors automobiles--Motors--Modification. 3. General Motors automobiles--Performance. I. Title.
 TL215.G4P67 2009
 629.25'04--dc22
                                    2009017864

Contents

Acknowledgments

Special thanks to:

Brian Nutter, Wiseco Pistons
Joe Hughes, Wiseco Pistons
Laura Johnson, Wiseco Pistons
George Rumore, Kooks Custom Headers
Randy Becker, Harland Sharp
Kurt Urban, Kurt Urban Performance
Nick Norris, Callies and Compstar
Steve Demirjian, Race Engine Development
Brian Tooley, Total Engine Airflow and Trick Flow Specialties
Jason Mangum, Precision Race Components
Tracy Dennis, Sunset RaceCraft
Georgia Crouch, Image Editing and Touchups

Introduction

The LS1 5.7-liter V-8 began production more than 10 years ago, but many people and even some engine shops still are intimidated by the Gen-III and -IV family of engines. Fact is, it's unlike any V-8 engine GM has released in the past. Even though it is no longer new, it's still quite unfamiliar to many.

The first LS1s were released the year after the limited-run LT4 debuted, which at that time was *the* modern marvel of the SBC world. In retrospect, the LT4 didn't last long before being dethroned. While the differences between the Gen-II LT1 and LT4 were mostly external compared to the traditional small-block 350-cubic-inch, the Gen-III used all new architecture and components. Therefore, there are no carryover parts between the Gen-III and its predecessors.

The all-new fifth-generation Corvette received the new LS1 engine package in 1997, while the F-body platform (Camaro and Firebird) got it the following year. The 1997 Corvette's 5.7-liter LS1 was rated at 345 horsepower, and the 1998 F-body 5.7-liter LS1 was conservatively rated at 305 horsepower. Ironically, there are minimal differences between the two, considering the huge horsepower rating difference. The horsepower rating was either underrated due to insurance issues, or it was thought that the Corvette was given the higher numbers to justify the $20,000 price difference. Since then, there have been many improvements and refinements to these engines. The 405-horsepower LS6s, 505-horsepower LS7s, and 638-horsepower LS9s are the most impressive advances. Trucks, SUVs, and some front-wheel-drive passenger cars even have variations of the LS-series engines.

The LS1 is much different from its elder small-block Chevy and LT1 siblings in all aspects. Despite the unfamiliarity and newness, it didn't take long for the performance world to take notice of the huge performance potential of this engine family. The LS1 has proven itself as one of the most responsive modern engines when it comes to performance modifications. It responds with glee to just about anything you throw at it, like a loyal pet given a treat.

The first few modifications came along fairly slow as there was a steep learning curve early on. There were a few homebrew modifications, such as ported throttle bodies, ported mass airflow sensors, and (gasp!) gutted catalytic converters, that woke up the engines well. Before long there were cold air intakes, under-drive crankshaft pulleys, and headers and exhaust systems. Then came camshafts, ported cylinder heads, intake manifolds, nitrous systems, turbos, blowers, and so on. It all complements the Gen-III and -IV engine family very well. There seems to be no wrong way to build one of these setups, although some choices do perform more efficiently than others.

You will find information related to the engines as a whole in this book, listed as "LS-series" throughout. Specific differences and changes per engine family will be isolated by specific name. The GM performance parts LSX block is listed as "GMPP LSX" when referring directly to that engine block. Also, when referring specifically to small or large power-adders, I would use the mindset of "small" being something the original equipment manufacturer (OEM) engine would handle. For N20 this would be 150 shot or less with the proper safety precautions and fueling. With boost, I would classify "small" as what the supercharger or turbocharger kit does in off-the-shelf form. If it produces 7 psi as supplied, that is what I would consider "small." Large power-adders are anything above these as-supplied basic numbers and would normally require greater attention to specific areas of the engine buildup for durability or longevity.

Nowadays, there are many choices to be made in how you wish to modify your LS-series engine setup. You can build all-out heads and cam setups on everything from a stock 5.7-liter short-block to monster displacement 454-plus-cubic-inch engines. The LS-series engine remains a developing performance architecture, and constant progress is being made, seemingly daily. In this book, you'll find information on many setups patterned to build your LS-series vehicle to your desired power level, from basic to extreme.

Chapter 1
Engine Block

The engine block is the foundation of any high-performance motor. It is the structure that holds everything together, but it has many other functions. Many people choose a block on the displacement sizing itself, but there are a slew of other factors that come into play when choosing the correct block to build upon.

Picking your engine block should be one of the first decisions made when building your desired engine. This isn't 1998. There are now an abundance of choices in the block arena, both from factory sources and the aftermarket. Factors such as displacement, intended application, weight, and block strength all come into play. The most important reason to make your block decision early on is that it's usually the first thing to be machined. You can change internal and external components later if you decide you made the wrong choice with the crankshaft, connecting rods, or pistons. Once the block is machined, however, you're pretty much stuck with what it is. It's something you want to get right the first time.

The LS-series family of blocks comes in a variety of materials and applications. Here, cast-iron blocks mingle with aluminum blocks while waiting to be bathed in honing oil.

DISSECTING THE GEN-III, A.K.A. THE "LS1"

The first Gen-III engines were the 5.7-liter aluminum-block engines known to the automotive crowd as the "LS1." To obtain this displacement, the bore and stroke dimensions are a 3.898-inch bore with a 3.622-inch stroke. This is a highly popular engine to modify because all 1997–2004 C5 Corvettes, 1998–2002 F-bodies, and 2004 GTOs used this basic engine, helping give the Gen-III the reputation it has today.

There were a few minor variations made to the block as the LS1 progressed. One change was from the LS1 block to the revised LS6 block. All Corvette Z06s received LS6 block, and certain 2002 F-bodies did as well. The main differences between them are block material and improved crankcase breathing, which was necessary due to increased engine rpm ranges.

The LS1 5.7-liter has made a name for itself on the street and has a reputation for slaughtering many Mustangs and Cobras. This is the bare, no-frills, aluminum LS1 5.7-liter block. Its reputation precedes itself.

These little stands are actually the locations for the engine knock sensors (if used). Many would call this the lifter valley, but if you notice, you cannot replace the lifters without removing the cylinder heads themselves. This is a minor inconvenience.

The change from long head bolts to the medium-length head bolts that were used throughout all Gen-III engine designations in 2003–2004 model years is not well known. All F-bodies used the longer head bolts stock, unless equipped with a replacement engine, as the last F-body was manufactured in 2002, long before the head bolt revision.

The 1997 and 1998 LS1 5.7–liter blocks had thinner cylinder sleeves, so one drawback of these engines is they cannot tolerate a lot of cylinder machining. The limit for these blocks is 0.005-inch overbore, while all other 5.7–liter blocks are allowed 0.010 inch.

Popular engine displacements using the 5.7–liter block for buildup are the 347-cubic-inch (3.905-inch bore x 3.622-inch stroke), 383-cubic-inch (3.905-inch bore x 4.000-inch stroke), and 395-cubic-inch (3.905-inch bore x 4.125-inch stroke). With the right combination, these little engines can put up quite a fight and make very respectable horsepower numbers. A vast majority of fast LSX-based cars are using the stock block, and quite possibly the same stock short-block the car came with. The LS1 is highly efficient and surprisingly durable in stock form. Big cubic-inch engine owners beware of these wolves in sheep clothing; power-adder 5.7–liter engines can embarrass you easily if set up correctly.

The aluminum block is poured around the cast sleeves, which have a small recommended machining limit compared to conventional iron blocks: 0.005 inch on 1998 blocks and 0.010 inch on 1999–2004 blocks. The lifters reside in plastic locating guides inboard the cylinders.

All LS-series engines come equipped with six-bolt main caps. The non-LS6 blocks used a gun-drilled hole above all mains that facilitated crankcase breathing.

Compare the LS6 block lower end's cylinder-to-cylinder breathing window to the non-LS6 5.7-liter block in the previous photo. This window facilitates higher rpm crankcase breathing. Remember that whatever your displacement is above the piston, that same amount of air is being thrown about in the crankcase as well.

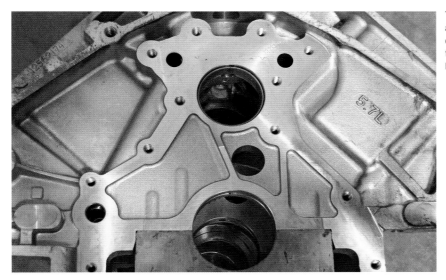

The oil travels from the oil filter directly to this point and makes a Y off toward each lifter oil gallery. The driver's side oil gallery is responsible for feeding the lifters, the camshaft bearings, and the complete bottom end of the engine.

Here is the newer design lifter oil galley. Notice this area is now opened up to provide unrestricted oil flow to both lifter banks (compare to preceding image). This update equalizes any discrepancies in oil flow that may show up under demanding conditions.

Not long after the world was introduced to the aluminum 5.7-liter, the redesigned 1999 Chevrolet Silverado and GMC Sierra trucks were released. These were available with three different Gen-III engine variants at 4.8-liter, 5.3-liter, and 6.0-liter displacements. Though none were particularly impressive in stock form, the reason for mentioning them is that Gen-III engines have many interchangeable parts that are desirable donors for certain buildups.

All of the Gen-III engines are fairly stout in stock form. The one drawback to the truck motors in stock form is that the pistons are cast aluminum versus the LS1 and LS6 pistons, which are of the stronger hypereutectic design. The 4.8-liter and 5.3-liter blocks are identical other than the rotating assembly. The 5.3-liter and 6.0-liter share the same crankshaft stroke like its older 5.7-liter cousin. All of these early truck engine packages were cast-iron blocks. Later, some trucks received aluminum 5.3-liter engines similar to the 5.7-liter aluminum block. One thing to note about the 4.8-liter and 5.3-liter blocks is the cylinder wall thickness. They have enough material to be bored out 0.120-inch to stock LS1 displacement. So with the 4.8-liter or 5.3-liter block, one could have the same engine displacement as the aluminum 5.7-liter but with a cast-iron block.

The 6.0-liter block is the most sought-after truck donor block. In stock form, the 6.0-liter iron block is one stout setup. This block is readily available, and it's the most economical to purchase if starting from scratch. It's equipped with a conventional 4.000-inch bore and can go up to a 4.060-inch oversize easily, though 4.030 inches is the maximum recommended bore size for a power-adder setup. Many of these blocks have been built to produce more than 1,000 horsepower to the wheels. Popular displacements for the 6.0-liter block are the stock 364-cubic-inch setup (4.000 x 3.622 inches), 370-cubic-inch (4.030 x 3.622 inches), 402-cubic-inch (4.000 x 3.622 inches), and the ever-popular 408-cubic-inch (4.030 x 4.000 inches). A 422-cubic-inch engine can be built taking bore and stroke to the limits, although it's unpopular to do so. Iron's strength comes with the cost of weight, as this block is about 85 pounds heavier than the aluminum block.

INTRODUCTION TO GEN IV

The 5.7-liter lived a good life, but 2004 was its last year in production vehicles. In 2005, the Gen-IV engine became available, making even more engine choices available. Besides the larger cylinder bores, not a whole lot changed, but it still pays to be informed.

The 2005 C6 Corvette and GTO were made available with the 400-horsepower LS2 engine. This closely resembles an LS6 engine design, but it uses a larger 4.000-inch bore. There are a few external differences with the most significant being that the camshaft sensor reluctor is moved from the rear of the camshaft to the upper timing gear itself. This necessitated a different front timing cover and timing gear. Camshaft and knock sensor harness wiring extensions are available for vehicles not originally equipped with a Gen IV engine.

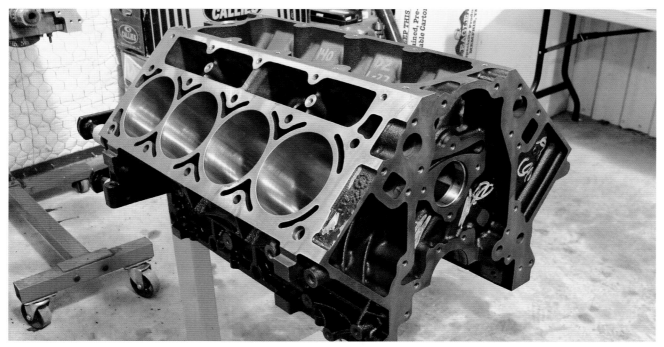

The iron LQ4/LQ9 is a stout foundation to build on. It has proven to hold up well beyond 1,000 rear wheel horsepower on numerous occasions. Standard bore is 4.000 inches, but it can reach a 4.060-inch bore for naturally aspirated applications; 0.030 inch is the recommended max for power-adders.

GEN III VS. GEN IV

Today, LSX engines can be had in a variety of OEM designations, engine sizes, and chassis setups. The move from the Gen III to Gen IV designation brought about some minor design changes that can be a nuisance. Most of what is changed isn't apparent to the untrained eye, but there are important changes throughout the entire engine that will significantly affect the retrofit of a Gen IV into a Gen III vehicle.

Comparing the last Gen III (LS6) to the first Gen IV (LS2), we will find that the obvious difference is the timing cover–mounted camshaft sensor. Although the sensor signal is identical, the reluctor is relocated to the upper timing gear rather than the rear of the camshaft itself. If putting the Gen IV–based engine into a vehicle with an LS1 or LS6, the camshaft position sensor harness must be extended, or a PnP extension harness must be purchased for the proper camshaft signal to reach the powertrain control module (PCM). Plug and play retrofit harnesses are available through the aftermarket to extend the cam sensor and knock sensor harness when converting from Gen III to Gen IV engines. Both Katech Performance and Caspers Electronics offer this kind of harness.

The Gen IV family is the first to receive active fuel management, which is a cylinder deactivation feature designed to achieve slightly better fuel economy. All Gen IV engines except the LS7 and GMPP LSX have this provision machined into the engine. The telltale sign is the presence of eight towers that reside under the valley plate. These are the oiling holes that, when turned on through the compatible DOD valley plate cover, shut off the lifter functions for that particular cylinder via electronic solenoids controlled by the PCM and diverting oil pressure into the lifter and disengaging it. These towers need to be plugged if not used to prevent an internal oil pressure leak.

Along with the active fuel management or displacement on demand, (DOD), the block has four identification tabs that locate the special lifters and lifter trays that these engines require in the corresponding cylinders.

Again, most performance applications do not use the special lifters, but because the block changes to the Gen IV are a "one-size-fits-all" design, the revised LS2-style lifter trays are required to clear these identification notches. The LS2 lifter trays can be used in either Gen III or Gen IV engine blocks. Gen III lifter trays can only be used in Gen III blocks.

Gen IV knock sensors are relocated to the external sides of the block. The Gen III knock sensors were located under the intake manifold in the valley plate well areas and were threaded into the centerline of the block. Gen IV knock sensors will not operate with Gen III electronics. If knock sensors are required for your build, you will need the OEM Gen III knock sensors and two unused M10 bolt hole locations such as on the side of the block. Locations depend on application, type of headers, and block access clearance. Note that relocating the knock sensors from the valley plate area to the side of the block may affect the operation and sensitivity of the knock sensor circuit, requiring custom tuning for proper operation.

Gen IV engines from 2005 have the same 24x crankshaft reluctor as the previous Gen III engines and the same 2x camshaft position sensor (just timing gear-mounted, as noted above). In 2006, certain Gen IV engines were upgraded to the 58x crankshaft sensor gear and 4x timing chain gear. This change offers better resolution to the PCM on the exact locations of these components. More accurate crankshaft and camshaft sensors were required for the variable valvetrain timing system (VVT). From a performance standpoint, VVT offers quite a bit of flexibility in the torque curve characteristics of the engine. Any camshaft has its optimum rpm range, and VVT can extend this range by a huge amount. The ability to advance camshaft timing at low engine speeds and then retard the camshaft timing for higher engine speeds opens an entire new door for camshaft design on Gen IV engines. Sadly, this is only a benefit that 58x crankshaft–equipped vehicles can use.

The aluminum-block 6.0-liter LS2, which the factory rated at 400 horsepower, is a stout engine in stock form. These were similar to the 5.7-liter LS6 in design, although they came with a conventional 4.000-inch bore size to increase displacement.

This block was the first to receive the active fuel management and displacement on demand (DOD) block machining provisions. The LS2 itself does not use DOD, but it requires using revised lifter plastic guide plates due to the cast locating tabs now included with Gen IV blocks and a special Gen IV valley plate cover. This is necessary to plug off the lifter bleed-down passages used to shut off cylinders. Similar engine sizing applies to the 6.0-liter LS2 as the 6.0-liter iron block: The stock 364-cubic-inch setup (4.000 x 3.622 inches) and 402-cubic-inch (4.000 x 3.622 inches) are the most popular.

It should be noted that the 6.0-liter truck engine is available in a Gen IV designation, identical to the LS2 features externally but made in cast iron versus the LS2's aluminum construction.

The Gen-IV lineage may have started with the LS2, but there have been a few other additions to the family. One is the 6.2-liter L92 and LS3 engines (4.065 x 3.622 inches). These blocks are visually identical to the 6.0-liter block, save for the larger bore size. The larger bore opens up some room for more cubes, allowing 416-cubic-inch, 418-cubic-inch, and 422-cubic-inch engines.

These odd-looking towers are part of the active fuel management or displacement on demand (DOD) lifter bleed-down holes (on all Gen IV engines). Not used on the LS2, these need to be plugged with the correct valley cover that includes O-ring seals to avoid internal oil leaks.

This close-up is the DOD lifter location and lifter towers that enable and disable this cylinder's lifters, when so equipped. The affected cylinders are 1, 4, 6, and 7. Using conventional lifters and plugging these lifter towers is a requirement for non-DOD engines when retrofitted to Gen III vehicles.

The 6.2-liter Gen-IV aluminum cylinder block (mostly known as the LS3 or L92) is found in certain SUVs and 2008 and newer Corvettes. It is similar to the LS9 engine block found in the 2009 Corvette ZR1 in terms of displacement dimensions. The 6.2-liter blocks have a 4.065-inch bore size standard.

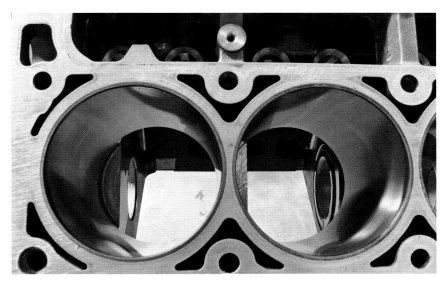

The LS3/L92 6.2-liter has many similarities to the 6.0-liter LS2. The only difference is a 0.065-inch increase in bore size. The 6.2-liters, like most Gen IV engines, come DOD-equipped also, but the system is left unused. SUV 6.2-liters come with variable valve timing (VVT) progressively migrated to other applications.

The LS3/L92 block can tolerate a 0.020-inch overbore easily. Some have even bored these to 4.100-inch bore sizing. When getting that aggressive, it's probably best to stay away from power-adders as the cylinder sleeve ends up quite thin.

Another major advantage of the L92/LS3 engine package is the cylinder heads. These are non-cathedral square-port heads, quite different from the Gen-III heads. Unported, these heads have huge airflow capabilities. See Chapter 3 (Cylinder Heads) for more information on the top end of this engine.

The truck version of the 6.2-liter, the L92, uses variable valvetrain timing (VVT) to produce a wider, smoother power band. Use of the VVT hasn't been widely explored in a performance application for LS-series engines, but camshaft manufacturers have been working with it and have shown some promising initial results. The supercharged ZR1 LS9 6.2-liter is also based on this block, albeit with a few changes to increase durability, such as dry-sump oiling and oil squirters to cool the pistons. The LS9 boasts an impressive 638-horsepower rating. When the factory is pumping out these kinds of numbers, you know that you've found a reliable setup.

One of the more recent high-profile engines is the 427-cubic-inch LS7 that is found in the 2006–2008 Z06 Corvettes with a 505-horsepower rating. Being a factory 427-cubic-inch, 7.0-liter engine, many may think there is little room for improvement. They are wrong. With a huge 4.125-inch factory bore size, this is a desirable block for building a large-cubic-inch engine. As delivered in the Z06, these are dry-sump engines, though this can be changed if using the block in another vehicle.

All conventional LS-series parts such as crankshafts, camshafts, and cylinder heads will bolt up as normal. The LS7 block uses pressed-in gray iron sleeves, unlike all other production aluminum blocks, which have in-block integral liners that are set in place before the block is cast. The factory LS7 sleeves need to be upgraded if the end user plans on running nitrous or any other huge power-adder. For naturally aspirated applications and small nitrous system use, the factory sleeves are adequate. The 430-cubic-inch and 440-cubic-inch are easy to attain with slight bore changes or more stroke.

Some other notable features of the factory LS7 short-block are the use of titanium connecting rods and hypereutectic aluminum pistons with valve reliefs. The hypereutectic aluminum pistons were used because forged pistons tend to have undesirable engine noise on cold startups. Piston slap is not something an average Corvette Z06 owner would like to hear from their new LS7 engine. There are several aftermarket piston forgings available for upgrading from the stock pistons.

LS7 engines have a 4.125-inch bore and 4.000-inch stroke as built, but the engines have a few drawbacks for all-out performance. Two main problems with the LS7 are the limitations of the hypereutectic pistons and the gray iron sleeves.

Gen IV 7.0-liter blocks are built on an aluminum base with dry sleeves added for the large bore size. Other than the large bore, these are similar in form to the other Gen IV blocks. Here is a view of the bay-to-bay breathing passages. Some rare race LS7 blocks do not use these breathing holes.

LS7 blocks are fortified in some areas more than others. The upgraded six-bolt main caps are one area not left behind. These are forged main caps with locating dowels. The dowels help keep the main caps from shifting around under load.

AFTERMARKET AND NONPRODUCTION, BUILDING FROM THE BEST

While aftermarket blocks are fairly new, GM Performance Parts (GMPP) had its own big-cube engine about six years earlier than the aftermarket. It was named the C5R and carried a hefty price tag that only the elite of the LS1 world could afford.

These were originally designed for Le Mans racing, but inevitably they were released by GMPP as a bare-block. The block was the basis for the newer LS7 block, so while there are many similarities, they are not identical twins. The C5R receives much more thorough machining attention and inspections than a production LS7 block would receive, plus there are differences in block, main cap, and sleeve materials. The C5R uses billet main caps with high-quality 4340 main cap and head stud fasteners.

One of the most obvious visual differences between the two is the camshaft sensor location. The C5R uses a conventional Gen-III location in the rear of the block, while the LS7 uses the newer timing cover–mounted camshaft sensor location. The only real downfall of the C5R block is its price, which at more than $6,000 is roughly twice that of the production LS7 piece. For those who desire high quality with meticulous machining craftsmanship, the C5R would not be a letdown.

GMPP again stepped up to the plate in 2007 to deliver a bulletproof block in the form of the "LSX" iron Bowtie block. Starting with a rough bore of 3.990 inches, this block can accommodate anything from a 4.000- to 4.250-inch finish bore size. That means it can be suited for a variety of displacements, as small as a stock 364-cubic-inch for the wasteful or easily as big as 454-cubic-inch for the greedy.

GMPP equipped this block with a six-bolt head, as well, but in a different fashion than the World Products block, meaning six-bolt heads do not interchange between the two blocks. Another major feature that deviates from the production-type blocks is the oiling system. It now feeds the main bearings separately before feeding the lifters. This is good news for solid-roller setups, as now one can easily restrict oiling to the top end of the engine and keep it with the rotating assembly where it's needed.

CHOOSING THE RIGHT BLOCK FOR YOUR APPLICATION

Look at your vehicle's usage and desired power levels to choose your engine's foundation. If you're drag racing a big-cube turbo setup, you would want the strongest and most durable block possible, with weight taking a back seat. If you are autocrossing the car, you would probably want a block with similar weight and attributes as the factory engine while keeping it durable. It's a game of compromises to narrow down your engine choices. Setting your goals early on and sticking to the plan is the best course of action.

STREET/TRACK DRAG RACING

Street cars are normally fueled with pump gas. Keep this in mind when planning your buildup, as when certain parts are subjected to detonation, things can get broken in a hurry. Depending on your intended usage, most street or strip cars should stick with OEM blocks and let power level goals determine if you should do an aluminum-based block or go with the tried-and-true cast-iron 6.0-liter block.

Of course, the desired displacement also plays a role. As a general guideline, most naturally aspirated and mild power-adder builds would be fine with any OEM aluminum block. If you're planning to use a large power-adder of some sort, then the iron block would be a wise choice. If the engine is overbuilt, it gives you room down the road for more power. While some have made more than 1,000 rear wheel horsepower with aluminum blocks, there are also people who don't get shot playing Russian roulette. Do it enough, and there will be a point when the odds catch up to you.

If your goal is to have a big-cube, pump-gas 427-cubic-inch engine, then you have to make a few other choices. This goal also limits your block choices quite a bit because of bore sizing. For something in the 600-rear-wheel-horsepower neighborhood, an LS7 block would fit the bill easily. If the scarce LS7 block cannot be procured, you can have an LS1, LS2, or LS6 block resleeved using Darton's Modular Integrated Deck (MID) sleeves, installed at a quality facility familiar with the sleeving process. The next option for a big-cube engine would be to use the GMPP LSX Bowtie Block, although it's quite a bit of an overkill for a true street or strip car.

A BLOCK THAT CAN STEP UP TO THE PLATE

You want to build a heavy-hitting power-adder car that only sees track duty and no or limited street use? In this case you want something beefy and reliable—a block that can step up to the plate and hit one out of the park every time.

For this application, at a minimum the 6.0-liter iron block should be used. The Darton MID sleeved block, GMPP LSX block, or the World Products Warhawk block would be a shoe-in for this type of strenuous racing activity. The aftermarket blocks use six-bolt cylinder head clamping systems that add additional fasteners to the deck and cylinder heads in order to keep the cylinder head gaskets in check under high boost or when spraying enough nitrous to put down an elephant. With fast power-adder cars, you want to do everything possible to keep the cylinder head gaskets intact, from keeping the tune safe to using thick-deck heads, running it pig rich, or using the six-bolt method. When used together, these measures complement each other well.

AUTOCROSS AND ROADRACING

The best choice for autocross and roadracing is to stick with an OEM aluminum block to minimize weight. The vast majority of autocrossers with LSX-based engines would be C5 and C6 Corvette drivers. The C5/C6 chassis is at home in this style of racing with minor chassis upgrades and has the oil pan design to supply the oil pickup tube during extreme cornering and braking maneuvers.

The blocks that would keep the weight similar to stock are, of course, all OEM blocks: LS1, LS2, LS3, LS6, and LS7, and the GMPP C5R block could be thrown into that mix. All of the OEM blocks hover around 110 pounds. One other option for larger displacements is to have an OEM Gen III block sent to Race Engine Development to have the Darton MID sleeving system installed, which can be installed into any OEM aluminum block. If you have a Gen IV block, Darton has a big-bore dry sleeve setup that is a little more economical.

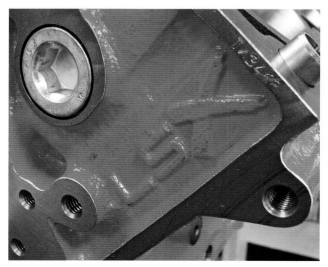

The GMPP LSX iron block is one stout setup. You automatically know that iron is stronger than aluminum, so this makes for the perfect choice for big power-adder engine setups that may exceed the capabilities of an aluminum block. The LSX block will support bore sizes of 4.000 to 4.250 inches.

The LSX block filled a void by creating an economical block package that uses six-bolt heads. The stock four-bolt locations are still used, with the addition of one extra bolt location above and below each cylinder. The four-bolt locations are stock 11-millimeter-diameter, while the extra holes use 8-millimeter diameter bolts.

The LS-series block comes with a little extra material on the deck. The deck height is 9.260 inches, which is 0.020 inch taller than stock. A clean-up decking is adequate, but if using off-the-shelf pistons, it will be necessary to deck the block 0.020 inch.

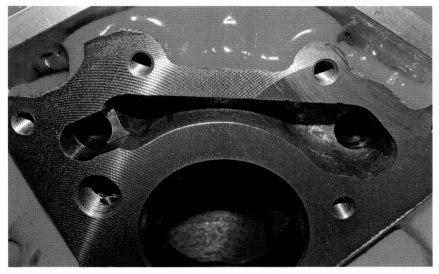

A key feature of the GMPP LS-series is the redesigned oiling system. It's designed as a priority system, which feeds the main bearings with oil first instead of the cam bearings. That 3/8 NPT hole is the main bearing supply.

The LS-series lifter galleries are fed oil that is supplied from the front of the crankshaft oil galley. Oil flow to the lifters passes behind the camshaft thrust plate and travels rearward, feeding all lifters. These two front lifter oil galley ports can be threaded and restricted for solid roller applications, as the bottom end of the engine is not fed from the lifter galleries in this block.

This oiling design requires a redesigned LS-series rear cover and camshaft thrust plate to divert oil as required. Luckily, GMPP thought ahead and supplies all that is needed for that system in the block crate. The six-bolt design requires different lifter trays that GMPP also thoughtfully includes. These lifter trays hold up fine to normal street abuse. For racing use, tie-bar lifters are recommended for strength.

Hot rodders tried resleeving the stock aluminum block quite early in the Gen-III's life. The first attempts at manufacturing big-bore LS-style engines seemed to be hit or miss. Some of these resleeved blocks would enjoy a normal engine life, yet many suffered from dropped or cracked sleeves and head gasket issues leading to massive coolant consumption.

Continued on page 25

Factory and OEM Blocks

Designation	Displacement	Material	Family	Bore Diameter	Max Bore	Siamesed	Misc.
4.8L/5.3L	4.8L/5.3L	Iron	Gen III	3.780"	3.910"	None	
4.8L/5.3L	4.8L/5.3L	Aluminum	Gen III	3.780"	3.910"	None	
4.8L/5.3L	4.8L/5.3L	Aluminum	Gen IV	3.780"	3.910"	None	DoD
LS1	5.7L	Aluminum	Gen III	3.898"	3.910"*	1/2 Siamese	
LS6	5.7L	Aluminum	Gen III	3.898"	3.910"	1/2 Siamese	Bay-to-Bay Breathing
LQ4/LQ9	6.0L	Iron	Gen III	4.000"	4.060"	None	
LQ9	6.0L	Iron	Gen IV	4.000"	4.060"	None	DoD
LS2	6.0L	Aluminum	Gen IV	4.000"	4.020"**	Yes	DoD
L76	6.0L	Aluminum	Gen IV	4.000"	4.020"**	Yes	DoD/No BtoB Breathing
L92	6.2L	Aluminum	Gen IV	4.065"	4.085"	Yes	DoD
LS3	6.2L	Aluminum	Gen IV	4.065"	4.085"	Yes	DoD
LS9	6.2L	Aluminum	Gen IV	4.065"	4.085"	Yes	DoD/Piston Oilers
LS7	7.0L	Aluminum	Gen IV	4.125"	4.130"	Yes	DoD - not drilled

Note: All OEM blocks are around the 9.240" deck height.
* 1997-1998 Blocks Maximum bore = 3.905"
** Factory Specification, although 4.030" bore obtainable

Aftermarket Blocks and Resleeved OEM Blocks

Designation	Displacement	Material	Family	Bore Diameter	Max Bore	Siamesed	Misc.
GMPP LSX	6.0L–7.4L	Iron	Gen IV	3.990"+	4.250"	Yes	9.260" Deck / 6-Bolt Headbolts
GMPP C5R	7.0L	Aluminum	Gen III	4.125"	4.160"	Yes	C5R Race Design
R.E.D. LS1*	7.0L	Aluminum	Gen III	4.125"**	4.200"	Yes***	Darton MID Sleeves
R.E.D. LS2-LS7	7.0L–7.2L	Aluminum	Gen IV	4.125"+	4.200"	Yes***	Darton MID Sleeves
R.E.D. LS2	6.0L–7.2L	Aluminum	Gen IV	4.125"+	4.190"	Yes	Darton Dry Sleeves
R.E.D. LS7	6.0L–7.2L	Aluminum	Gen IV	4.125"+	4.190"	Yes	Darton Dry Sleeves
R.E.D. C5R	6.0L–7.2L	Aluminum	Gen III	4.125"+	4.190"	Yes	Darton Dry Sleeves
World Warhawk	6.0L–7.4L	Aluminum	Gen III	3.990"+	4.155"	Yes	9.240-9.800" Deck / 6-Bolt

* LS6 Blocks can be used, but due to bulkhead windowing for bay-to-bay breathing tend to be weaker.
** Two Sleeve sizes available: 4.125"-4.160" or 4.170"-4.200"
*** MID Sleeves are butted against each other but feature coolant passages between sleeves.

RESLEEVED LS1–LS6 BLOCKS: PROBLEMS AND CURES

Since the introduction of the LS family of engines, people have been striving to build bigger ones. As the factory LS1 engine and all-aluminum LSX-based engines are sleeved from the factory, you would think this might be a rather simple process: remove sleeve, install new sleeve, and rejoice! That was never the case with early attempts at resleeving LS blocks. Usually the owner of the car would receive his or her new engine with a slew of headaches to go along with it. While some resleeves actually worked, many more failed, with dropped sleeves and, more commonly, coolant and oil-consumption issues.

The LS1 and LS6 were the only OEM aluminum blocks available at this time. These blocks are non-Siamese blocks and have a coolant passage between the cylinders at the top half of each bank. Current Gen IV aluminum blocks can be dry sleeved using Darton's Seal Tite dry liners as discussed in the "Dry Sleeving" sidebar because they have Siamese cylinders and do not have extra cooling between the cylinders.

These shops used sleeves with an outside diameter of 4.2500 inches. When bored to the common size of 4.1250 inches, this left a sleeve wall of only 0.0625-thickness. Think of a spark plug gap and add a sheet of paper to compare.

When boring the LS1 or LS6 block to accommodate the 4.2500-inch outside diameter of these sleeves, the boring procedure either hits the coolant passage or leaves the remaining supporting aluminum paper-thin, weak, and prone to cracking. The result is a coolant leak into the crankcase sooner or later. If by good luck you beat the odds on coolant seepage into the crankcase, the bad thing is you still lose: Dry sleeve liners require sufficient supporting material around the circumference of the sleeve to keep the cylinder round. The lack of material in the LS1 and LS6 blocks makes this impossible, guaranteeing oil consumption, blowby, and general lack of power, leading ultimately to the engine's demise.

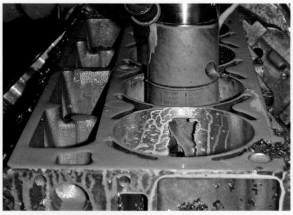

Race Engine Development starts the MID resleeving process by removing the factory iron liners. Shown for reference is the LS1 block. Notice the coolant passage between the cylinders. This coolant passage is the primary reason LS1 and LS6 blocks cannot have big-bore dry sleeves. *Race Engine Development*

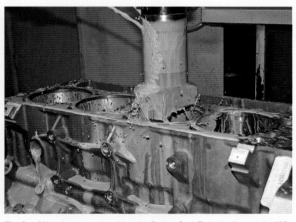

The Gen IV block can receive either the Darton Seal-Tight dry liner or the MID wet-sleeve system. The first process with any aluminum block is to remove the gray iron factory liner completely. *Race Engine Development*

Once the liner is removed, the factory aluminum cylinder wall is removed to just above where the MID sleeves will seat. The use of coolant while cutting keeps unwanted heat out of the block. *Race Engine Development*

Quite a lot of material is removed during the process. This is the empty cylinder case, devoid of the OEM sleeve and aluminum backing material. *Race Engine Development*

The block is then removed from the CNC machine, and vibratory stress is relieved. The stress caused by the extensive machining process causes the block to distort the first time it's fired up if it hasn't been stress relieved. *Race Engine Development*

After returning to the CNC machine, the bottom bore diameter where the sleeve O-rings seal is precisely bored to a proprietary tolerance. The fit is +/- 0.0002-inch maximum on diameter. *Race Engine Development*

If you have been around the LS community long enough, you probably have heard about these resleeved engines dropping sleeves. This causes huge losses in engine performance and coolant leaks that originate at the head gasket and the support structure behind the sleeve when the sleeve breaks free. While all sleeves have a lip at the top to prevent movement, when mated to a soft aluminum block this lip needs to be resting on the mating block step with a minimum of 0.100 inch of material overlapping, and 0.125 inch is preferred. If we go back and look at the sleeve dimensions mentioned above and calculate the measurements of the full-round dry liners, we will find that this seating ledge was only approximately 0.065 inch in the best case. If you have one of these blocks with no issues, go buy a lottery ticket and reap your good luck to its fullest.

The only proper way (and the preferred method) to have a large-bore resleeved block based on the LS1 and LS6 block is to use the Darton Modular Integrated Deck (MID) liner system. The MID is versatile enough to use on an Iron LQ4/LQ9 6.0-liter truck block for even more bottom-end strength. The MID liners are a thick wet-sleeve design. If you are familiar with diesel truck engines, they are quite similar to the serviceable sleeve liners that are in almost all diesel trucks on the road today. With each sleeve integrated against the next, the strength of the system as a whole drastically increases. The MID Gen IV blocks required a different sleeve than the Gen III aluminum blocks because GM revised the coolant floor, which changes where the bottom of the sleeve seals.

The MID wet-sleeve system is encircled by engine coolant. Evans Type R anhydrous coolant is a recommended upgrade with the MID blocks to prevent corrosion of the new sleeves and to provide more reliable cooling to the engine. The sleeve design promotes improved cooling with coolant passages between mating sleeves and around the perimeter of each cylinder through the deck surface. The new Gen IV LS2–LS7 MID sleeve is externally phosphate-coated to prevent rust if the engine is run on an engine dyno with plain water. One huge advantage of the MID sleeves is the ability to replace a damaged sleeve if needed.

The sleeve seat is machined above the O-ring location. Cylinder no. 1's location is flat machined to avoid cutting through the front of the block. The no. 1 sleeve has the matching flat to coincide. *Race Engine Development*

The LS1–LS6 MID sleeve comes in two part numbers and size ranges, 4.125-inch to 4.160-inch bore sizing and 4.170-inch to 4.200-inch bore sizing. The LS2–LS7 MID sleeve covers the entire "big-bore" range of 4.125-inch to 4.200-inch bore sizes with one part number. The sleeve length is 5.800 inches, which accommodates longer stroke crankshafts and adds much better piston support than the OEM sleeve lengths.

Next, the top bore diameter that locates the top diameter of the MID sleeves is machined via circular interpolation. The bore-to-bore cylinder spacing of 4.4000 inches is kept to +/- 0.0005 inch tolerance. *Race Engine Development*

The LS2/LS7 sleeve is phosphate-coated for use with plain water if run on an engine dyno. The Darton sleeve is similar to diesel sleeves in design. The grooves close to the top half are cooling fins to help cooling system efficiency. *Race Engine Development*

Once the block is machined for the MID sleeves, the sleeves can be prepared for assembly. This involves installing the three sealing O-rings at the bottom of the sleeve and coating them with a specific O-ring lubricant. *Race Engine Development*

The factory sleeves are first removed completely from the core block by boring out the original sleeves completely. Then the supporting aluminum cylinder wall that remains after boring is removed to just above where the MID sleeves end up seating. Once the bulk of material removal is finished, the block is set up on a vibratory stress relief machine, which corrects the distortion caused by the stresses imparted by the machining processes. A full-race block will be vibratory stress-relieved once more after the sleeves are installed, before final machine work is performed. After the initial stress relief, the block is set up on the CNC machine and prepped for sleeve installation by a series of machine work steps that clear and size the block for the matching sleeves. The block is deburred and washed before installing the sleeves. The sleeves are installed using O-ring lubricant on the three sealing O-rings at the lower sealing surface of the sleeve. Loctite 515 or 518 is used to effectively secure the top of the sleeve into place.

Once the flats are aligned and sleeves carefully installed and fully seated into place, a deck plate is torqued down using head gaskets to make sure the sleeves are fully seated. The Loctite needs to set up overnight before removing the torque plates. After decking is performed to ensure a flat deck surface, the block is bored out and then honed to size, much like any other normal block.

Loctite is applied to the upper sleeve locating flange, and the sleeves are inserted one-by-one on each engine bank in proper order. *Race Engine Development*

With all sleeves in their proper locations, they can be fully seated into the block. The cooling holes you see at the tops of the sleeve help remove heat from the cylinders. More holes are located on the exhaust sides than on the intake side. *Race Engine Development*

Once both banks of sleeves are in place, they are clamped into place by torqueing on two deck plates and letting the Loctite set up overnight. *Race Engine Development*

Once the Loctite has set, the block is ready to be machined. The tops of the sleeves first need to be machined flat by resurfacing the deck of the block on each bank. After decking, normal machine shop operations resume, and the block can be bored out to size. *Race Engine Development*

RESLEEVED GEN IV BLOCKS, MID ALTERNATIVE

One nice thing about the Gen IV blocks is that these blocks are tolerant to having dry sleeves installed to gain larger bore sizes, whereas Gen III blocks are not. Gen IVs can use either dry sleeves or the MID sleeves. The stronger sleeve is the MID sleeve, but the Darton Seal Tight "LS7" dry liner holds its own in the power department.

To install these dry liners into an LS7, the pressed-in factory liners must be removed first. If the liners are being replaced because of a failure and the supporting aluminum behind the sleeve is damaged, this must be repaired first. If the sleeve has insufficient or cracked backing material, coolant will wick its way past the crack and sleeve into the crankcase.

When installing these liners into the typical Gen IV aluminum block, the OEM sleeves must be machined out and the block machined for the dry liners. This involves machining the bore size correctly and then the sleeve seating upper ledge must be machined to proper diameter. The sleeves have a flat on each upper flange where it would otherwise overlap the next cylinder's sleeve. This flat is necessary to have the maximum amount of seat width due to the extra sleeve flange diameter. The dry sleeve has a seating ledge of 0.125 inch around the perimeter other than the intersecting sleeve flats on the flange. The minimum required to prevent the sleeve from sinking into the soft aluminum is a 0.100-inch step, with the Darton Seal Tight easily covering the minimum. There is a zero failure rate with this design when installed by a competent facility.

Dry sleeving has its advantages and place, and it is more economical than the MID sleeving process. The upgraded dry liners can reach horsepower levels well in excess of what is capable with a naturally aspirated or average power-adder engine.

One of the big advantages to using the Gen IV block is that it is readily available just about from anywhere, and for much less than the LS7 block, which is scarce and expensive to procure.

The LS7 and C5R blocks are great blocks until you use them for something they were never intended for. Here someone detonated an engine and cracked and shattered the upper flange off the sleeve. This incident also cracked the aluminum behind the sleeve. *Race Engine Development*

To install the Darton Seal-Tite sleeves, the upper flange must be enlarged a corresponding amount. The Darton sleeves use an abundant upper flange so thatßß the sleeve does not budge in the soft aluminum. *Race Engine Development*

The Darton Seal-Tite sleeve is manufactured from ductile iron, which is approximately three times stronger than the factory LS7 sleeve. This is what all LS7 engines should come with stock. *Race Engine Development*

After installing the Darton dry liners, the block must be decked to ensure a flat sealing surface and to even up any irregularities in sleeve manufacturing height. *Race Engine Development*

One company that pursued a better method for re-sleeving blocks was Race Engine Development (RED), working with Darton Sleeves. RED and Darton came up with a new sleeving design named "Modular Integrated Deck" (MID), which ties together all the sleeves on each bank of cylinders to increase strength and reliability. The process involves removing the cast-in-place iron sleeves and most of the supporting aluminum sleeve backing and then replaces them with high-quality wet-design sleeves that can be bored to a variety of sizes. This process can be done to all aluminum-block LS-series engines, and RED has also sleeved iron LQ4/LQ9 blocks using the MID sleeves with much success.

When using LS7 core blocks, a similar dry-sleeving process as the factory design is performed but with a much stronger sleeve material. Darton's sleeve material is approximately three times stronger than that of the stock LS7 sleeves, which are known to fail under detonation. Darton sleeves are made one at a time from centrifugally cast ductile iron, which will bend before breaking or cracking, unlike the stock gray iron sleeves.

FURTHER BLOCK THOUGHTS

Once you have chosen your block, the next step is finding a competent machine shop to perform machine work and, possibly, assembly. Even though the Gen-III and -IV engines have been on the market for quite a while, you may find it to be a chore to find a machine shop with the familiarity required to machine your new block to the exacting specifications required.

It is important to choose a machine shop that is experienced with the LS-series engines. Aluminum blocks have specific requirements for clearances and other nuances that only experienced shops will understand. If your chosen shop has never touched an LS-series style engine, you probably don't want them to use *your* buildup as a guinea pig. Aluminum blocks expand substantially when subjected to machine work operations. A skilled machinist takes this expansion into effect when machining the block so that when the block is room temperature again, the clearances are spot on. Their work may be sufficient for a stock engine buildup, but with anything 400 horsepower and above, quality machine work is required if you want your new investment to live longer than 10,000 miles. Choose your machine shop based on prior success and ask many questions pertaining to the build.

Chapter 2
Rotating Assembly

At this point you should have a good idea of which direction you are heading with your new engine. You've chosen the type of block, and the displacement is determined. The rotating assembly contains another set of critical choices to be made. This assembly consists of all moving parts in the bottom end of the engine and is the backbone of the engine. Typically the weakest links in the stock rotating assembly are the stock connecting rods and pistons, as is the case with the stock LS1 engines. A lot of power can be put through the stock short-block, but there is a limit to the amount of abuse it will take. When building a new engine, you have the opportunity to upgrade needed parts from a clean slate, and you do have a few choices.

Along with the matching engine block, the rotating assembly determines engine size, as the piston and crankshaft help determine the engine's displacement by a calculation of the bore diameter and stroke length. The piston and crankshaft dimensions determine bore and stroke, and the volume in the piston is a big part in getting the target compression ratio.

Rotating systems can be pieced together and balanced at your local machine shop. For those not privileged with a nearby quality machine shop, you can purchase a pre-balanced rotating assembly from the manufacturer or shop of your choice. It may cost extra, but buying a balanced rotating assembly in kit form can save you time and aggravation in your engine build.

When aimlessly wandering around the Callies/Compstar warehouse, one may become lost in a forest of crankshafts. Callies/Compstar offers premium, top-of-the-line crankshafts, budget-oriented forged crankshafts, and even connecting rods in a variety of flavors without compromising strength.

CRANKSHAFTS

The stock LS1 crankshaft is a nodular iron casting. All factory performance applications are gun drilled through the center to reduce reciprocating weight. The stock LS1 crankshaft stroke is 3.622 inches, which is identical to all other LS-based engines other than the 4.8-liter truck engine (3.250-inch stroke) and the LS7 427-cubic-inch engine (4.000-inch stroke). These stock LS1 crankshafts have proven highly reliable in the 600-rear-wheel-horsepower range in daily driver vehicles, with quite a few making several hundred horsepower more than that. Anyone building an engine with high triple-digit horsepower numbers should seriously consider upgrading to a forged-steel crankshaft as insurance against engine failure. That being said, there are quite a few LS1 fanatics that continue to roll the dice with big power stock short-blocks, and they are often successful. Many aftermarket crankshaft manufacturers have produced forged crankshafts for stock cubic inch fanatics. The upgrade to one of the forged stock stroke crankshafts will make for a virtually bulletproof setup.

Stock LS-series crankshafts have shown time and again that they are a stout piece in the factory engines. Few engine failures are attributed to the stock crankshaft giving up. The factory stroke on all OEM LS-series crankshafts is 3.622 inches, other than the 4.8-liter and 7.0-liter crankshafts.

The rotating assembly consists of the components directly connected to the crankshaft. The crankshaft, connecting rods, pistons/rings, wrist pins, and engine bearings all make up the rotating assembly. This is a balanced rotating assembly; we cannot change one major part without rebalancing the entire setup.

If your LS1 engine threw a rod or spun a rod bearing, it likely trashed the crankshaft also. If the damage is caught early enough, the crank may be salvageable by a competent machine shop that can regrind the journal undersize. If this is the case, matching undersized connecting-rod and main bearings would be necessary. These are widely available in the aftermarket. Note that unlike cylinder bores, which are machined oversized, there is no choice but to machine the crankshaft undersized to reveal a new wear surface. If ground incorrectly, this may significantly weaken the crankshaft. When a crankshaft is ground undersized, thicker bearings must be used to compensate for the material removed.

When you build a new engine, you may not want to spend a ton of money to have exactly what you had before, when for a few dollars more you can build something significantly larger. Thankfully, the LS-series aftermarket has provided well for those of us who like to build big engines, often called "strokers."

Many old-school engine builders will relate to the small-block Chevy (SBC) or LT1 383-cubic-inch when you are discussing strokers. These were and still are respectable and popular engines to build out of those blocks. It's a little different in the LS-series world, as you can still build a 383-cubic-inch stroker, but it uses entirely different dimensions than what a Gen I SBC builder would think of as a 383-cubic-inch stroker. To build one from an LS1 block, a 4.000-inch crankshaft is needed (a full 1/4 inch more stroke than SBC 383-cubic-inch engines).

The OEM crankshaft on the right is nodular iron, while the Eagle crankshaft is forged 4340 steel.

The Eagle 4.000-inch stroke crankshaft is ESP Armor coated for higher wear on the bearing surfaces, reduced friction, and increased oil shedding because of the shiny, polished finish.

This Eagle crankshaft is a 4.100-inch stroke crankshaft. The engine builder should verify part numbers and make sure the components match each other in the rotating assembly before going any further with balancing or assembly.

The rear of the 4.000-inch crankshaft is shown here. The rear main seal flange is the same depth as the factory crank, making engine swaps and buildups between LS-series-based vehicles much easier for the end user.

There are many manufacturers of LS-series crankshafts, with a variety of stroke sizes to choose from. One of the most popular and common sizes for various applications is the 4.000-inch stroke crankshaft, followed by 4.100-inch, 4.125-inch, and the forged 3.622-inch crankshaft, in order of consumer popularity. Aftermarket crankshafts are made many times stronger than the stock crankshaft, due to their material, forging, and machining processes. All of the aftermarket crankshaft manufacturers have their own designs, features, and tweaks. Many use the common forged 4340 steel material, with a few using billet steel. These cranks are cut out from a solid blank of metal. Further, certain crankshafts are then hardened by a nitriding process that hardens the outer layer of the crankshaft further under the outer surface by a minute amount. This hardening can be ground off. If remachined, the crankshaft will have to be rehardened to regain the same strength it had before grinding.

All of the main and rod pins will have 0.125- to 0.140-inch radii (which varies with manufacturer) where the main and rod journal wear surfaces meet the crank counterweights. This radius helps prevent stress points in the crankshaft and distributes loads more evenly than a straight-cut journal. The use of radii journals requires the use of chamfered or narrowed rod bearings to prevent the bearing edge from binding on this radius surface.

It may be hard to tell in pictures, but there is a small 0.125-inch radius where the connecting rod pin meets the webbing of the crankshaft itself. The radius distributes load better than a straight cut, which is prone to creating a higher stress point. The chamfered bearing oiling holes are visible in this image.

Here's another view of the rod pin where the radius can be viewed. The Eagle and many other aftermarket crankshafts also have lightened rod pins, which helps to remove reciprocating weight from each crank throw. Less weight with the same strength is always better for moving parts.

Now is a good time to compare rod bearings. The shinier front bearing is a stock-width replacement. The gray-coated bearing is a narrowed bearing designed to avoid bearing interference with the radius on the crankshaft's connecting rod journal. A narrowed or chamfered bearing must be used with all non-OEM crankshafts.

Crankshaft snouts can be procured with the standard snout length that works for most LSXs, but many offer the extended snout for use with the factory dry sump design. Price varies from the $800 range to $2,000 for the 4340 crankshafts, with cost determined by material, machining costs, and material source. Many crankshafts are forged overseas, with final machining and quality control done by experienced crankshaft grinders in the United States before being shipped to LS-series engine builders and parts warehouses. This does not seem to affect the engine longevity characteristics for 99 percent or more of the end users. This would be more of a personal preference or budget-oriented decision. Choosing a crankshaft completely manufactured and machined in the United States adds significantly to the price.

Even though some crankshafts are forged and machined overseas, proper clearance tolerances must be maintained. Here a Callies/Compstar crankshaft technician inspects journal sizing with a pneumatic rod pin tester. Proper quality control at this level weeds out any problems that may arise.

Viewing the standard-length crankshaft snouts, you can spot the crankshaft keyway slot. Eagle cranks do not come with a keyway and use the standard LS-series crankshaft key from GM. The GM keyways often need to be narrowed through filing or sanding on the keyway sides to fit into the Eagle keyway slot. The GM keyway is 0.010 inch wider than the Eagle keyway slot.

LS-series crankshafts use a pressed-on reluctor wheel to tell the powertrain control module (PCM) about its location. The main data the PCM needs from the reluctor wheel is when the pistons reach top dead center (TDC) on cylinders 1 and 6. These two cylinders are at TDC at the same time. The camshaft sensor signal determines where in the firing order these two cylinders are, as the camshaft rotates once for two crankshaft revolutions. All Gen III and early Gen IV engines used a 24x signal for crankshaft signals and then the mating 2x camshaft signal, which is exactly enough input for the computer (PCM) to determine the start of the firing order after just one crankshaft revolution. During 2006, GM changed from the 24x crank signal to the newer 58x crank signal, improving crankshaft position resolution. GM also

The rear of the crankshaft is the location of the crankshaft reluctor wheel. This particular crankshaft has the 24x reluctor wheel, meaning it will eventually be used in a 1997–2005 Gen III vehicle or a 2005 Gen IV vehicle.

This 4.100-inch stroke crankshaft has the newer 58x reluctor wheel, which is intended for 2006 to 2009 and newer Gen IV-based LS-series engines. You cannot use this in any Gen III vehicle and only in a few early Gen IV vehicles. Most 2006 and newer cars and trucks use the 58x wheel.

With the crankshafts situated side by side, it is easy to spot the difference in reluctor wheels. This is an important thing to pay attention to. If you match the reluctor to the wrong vehicle, you will be in for a nasty surprise when you attempt to start your new engine.

Here is a clean visual view of the correct reluctor wheel alignment. With the rear flange dowel hole at 12 o'clock, the two alignment triangles/semicircles need to face the rear *and* point to the 9 and 3 o'clock positions. Check this well before balancing. Every crankshaft manufacturer other than GM has had problems with this alignment it seems.

changed to a 4x camshaft signal (there's more on that hardware in the camshaft section, Chapter 4). Crankshaft reluctor wheel design is not a choice you may make: You *have* to use the design that your vehicle and PCM originally had. You cannot use the 58x reluctor in 24x vehicles and vice versa.

If you have one of the few in-between year vehicles (2006) and are unsure of which setup you have, GM made them easy to identify with color-coded crankshaft and camshaft sensors. Black sensors indicated 24x crankshaft reluctors, and gray sensors indicated a 58x reluctor wheel. Verify that you have the correct reluctor wheel before balancing your rotating assembly. This is one of the things you do not want to find out is wrong when your new engine does not start up after spending hours or even days installing it.

The Compstar-forged 4340 crankshaft is nitrided for surface strength and long bearing wear. Compstar offers either reluctor wheel for exact application fit and offers the LS7 front snout to cover all applications. Each crankshaft comes rough pre-balanced to 1,750 grams bob weight. *Callies/Compstar*

Compstar 4340 crankshafts have to meet an exacting standard of tolerances before being shipped out. All crankshafts are Magnafluxed to check for cracks or irregularities and checked for runout, and the stroke is measured. *Callies/Compstar*

If light weight and strength are what you want, the Callies Magnum XL has it. Material is removed in noncritical areas to reduce reciprocating weight and to effectively alleviate oil windage. The rod and main journals are gun drilled for weight savings, the crank is balanced to your bob weights, and it's all strengthened with a premium heat treatment. *Callies/Compstar*

CONNECTING RODS

Connecting rods are an integral component of the rotating assembly, and you do not want to skimp on them. The stock connecting rods are made of powdered metal and have a cracked-cap design, which is reliable and holds its own in the power capacity department. The stock LS-series connecting rods are 6.098 inches long (not including stock 4.8-liter, LS7, and LS9) and have an oddball 0.945-inch piston pin size, which makes them incompatible with most aftermarket pistons and crankshafts. For the relatively small expense aftermarket 4340 forged rods represent, there is really no reason to use stock connecting rods unless you are building a stock piston or rebuilding a stock-based engine. When you factor in the price of upgrading the connecting-rod bolts to the much stronger Automotive Racing Products (ARP) cap screws, resizing the big end of the connecting rod, and then pressing on your new pistons, the cost difference to upgrade to aftermarket connecting rods is minimal, and in some cases it may be more economical to upgrade instead. If reusing the stock connecting rods, a recommended upgrade is to replace the connecting-rod bolts. ARP offers 8740 and ARP 2000 bolt materials as an upgrade to the stock fasteners.

The stock LS-series connecting rod is a stout piece for a factory item. The rods are an I-beam design and are manufactured from powdered metal. This rod is the newer full-floating design that uses c-clips to hold the piston pin into place, although most Gen III LS-series engines use press-fit pins.

The stock connecting rod uses a unique alignment method. The connecting rod is made in a one-piece design, and a small stress line is machined in the connecting rod end cap. The rod is then subject to strain, which cracks apart the connecting rod cap from the rod itself.

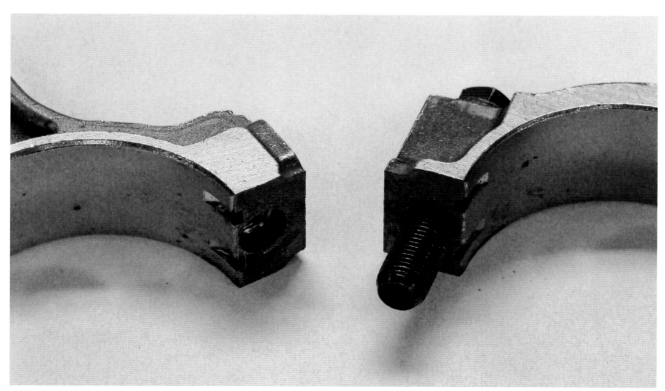

This design ensures that the rod caps line up perfectly every time, as each fracture crack will only pair up correctly to its matching mate. Don't lose one or you'll have to buy a new connecting rod. When tearing down an LS-series engine with these rods, it is helpful to scribe the rod numbers on both the rods and the rod caps so as not to mismatch them.

To understand which connecting rods are needed in your application, it is beneficial to understand why connecting rods fail. Pure power alone will seldom break a connecting rod in half. It will likely bend before breaking by overpowering the material strength. Connecting rods break under the "stretching" force generated by pulling the piston back down the cylinder from top dead center. The piston is slung away from the crankshaft on the upward stroke and then pulled back suddenly in a never-ending cycle. The stress worsens with longer stroke and more rpm.

Connecting-rod bolts are also subject to this same pulling force, but the rod bolts also have the connecting-rod weight to support on top of the piston weight. This is one big reason to keep high-rpm engine components as light as possible while retaining strength and durability. You can imagine how much force a heavier piston places on a connecting rod versus a lightweight piston at 7,000 rpm. As an example, a standard-build 427-cubic-inch LS-series (4.125x4.000 inches) turning 7,000 rpm has an effective stress of slightly more than 10 tons on the connecting-rod bolts. If you are building a high-rpm screamer, do not skimp on connecting rods.

There are several varieties of aftermarket rods. There are I-beams, H-beams, aluminum alloys, titanium, and billet steel connecting rods. Some are much lighter than others, especially so with the aluminum and titanium alloys compared to forged steel rods. The benefit of a lightweight rod is best seen when accelerating the engine into the high-rpm bands, as less rotating weight allows the engine to rev faster.

The most common aftermarket connecting-rod length for LS-series engines is 6.125 inches, which is 0.025 to 0.027 inch longer than the stock connecting rods. A few manufacturers had manufactured stock-length forged rods in the 6.100 inches length, but they ended up being unpopular.

I-beams tend to be the lightest steel rods of the group, but even they are usually much stronger than stock connecting rods. From a weight standpoint, I-beams are lighter than H-beams by design and shape. The I-beams resemble the stock connecting-rod design but use a stronger 5140 or 4340 material and stronger, larger-diameter connecting-rod bolts from ARP of 180,000-psi tensile strength 8740 material, or optional 220,000-psi ARP 2000 cap screws. These are a good choice for stock-type buildups to mild buildups and even minor power-adders but are not recommended for large power-adder engines in basic form. Most would expect the budget-oriented I-beam rods to handle 600 horsepower in a carefully built setup, with some high-end I-beam rod designs rated up to 900 horsepower. Certain billet I-beams are rated for up to 1,100 horsepower and carry the price tag to go with it.

The 4340 I-beam connecting rod has a similar profile to a stock connecting rod, but it is equipped with stronger rod cap screws and much improved rod material. All aftermarket rods will have a chamfer on the big end of the rod. This is to keep the rod from seizing on the crankshaft radius flange. Stock rods do not have this chamfer. *Callies/Compstar*

All Compstar connecting rods have premium features without breaking the bank. The Compstar rods come standard with Automotive Racing Products (ARP) 2000 cap screws and are weight matched within 2 grams of each other per set. A sample rod is pulled from each set of eight and checked for center-to-center length, rod parallelism, and bend/twist. *Callies/Compstar*

The popular 4340 H-beam design is available from numerous manufacturers. Shown here is the Compstar 6.125-inch LS1-specific connecting rod. The H-beam rods are physically wider than the I-beam throughout the length of the connecting rod, but the "H" portion is machined out for weight savings while keeping the strength of the connecting rod intact. *Callies/Compstar*

The H-I Beam rod is a new spin on the H-beam design, combining features from both the H and I beams to make a new, stronger line of connecting rods. The beam's ridge thickness is increased from 0.082 to 0.110 inch, an increase of 25 percent over the standard H-beam, while only increasing weight 6 percent with an I-beam-style side profile. *Callies/Compstar*

The Compstar H-I beam rods are intended for power-adder heavy hitters, using extra material to increase strength in high-strain areas of the connecting rod. Further, the H-I beam is designed with extra material in the wrist pin and big end sides of the rod to increase capacity. These are considered a step above the original H-beam rods. *Callies/Compstar*

The 4340 H-beam rods are the most popular with the LS1 aftermarket. The H-beam design is physically stronger than the I-beam design when comparing similar materials, although the benefit isn't fully realized until you exceed the power capacity of the I-beam. H-beams with 7/16-inch-diameter 8740 cap-screw bolts are rated at 700 horsepower, although that has been easily surpassed with numerous builds. For those over that rating, the upgrade to ARP 2000 or L-19 rod bolts is highly recommended. Rod bolt upgrades make it possible to almost double the horsepower capability rating of the connecting rod. Users of nitrous or any power-adder setup should definitely consider the rod bolt upgrade when ordering connecting rods; overkill with rods or rod bolts will not have any detrimental effects. The upgraded bolts can be bought separately if you already own connecting rods with 8740 cap screws, but the big end (bearing side) of the rod should be rehoned, as changing rod bolts or torque settings affects rod cap distortion to an extent. Verifying roundness distortion should be done as a minimum when changing rod bolt materials and torque.

The Callies/Compstar line of connecting rods come weight matched to each other. Rods are grouped by big end weight first. Then sets of eight are matched by small end weight to within 2 grams. This keeps each set of eight rods balanced as well as reasonably can be.

Connecting rod bolts are offered in a few different strengths. ARP 8740 is the standard bolt in many connecting rods and is good for the 700- to 750-horsepower range. Upgraded ARP 2000 and L-19 materials are rated for 900- to 1,500-horsepower engines. The stronger rod bolts can be used in any build. You don't need 1,000 horsepower to justify using these upgraded fasteners.

ENGINE BEARING CLEARANCES

All bearing clearances, both crankshaft and camshaft, have a direct effect on engine durability and oil pressure. Ask different engine builders what bearing clearances you should run and you may get a few different answers, all of which may work. The cast-iron 6.0-liter and GMPP LSX do not require anything special compared to other cast-iron engines, while the aluminum alloy block expands much more with heat, so aluminum block build clearances always need greater attention to avoid oil pressure issues.

Ideal bearing clearances change depending on usage, operating conditions, and even oil selection.

Although we set bearing clearances when the engine is cold and the clearances differ from each block material used, the clearances that matter are what the engine sees when it is running and warmed up. The solution is to build it tight on the engine stand and calculate the "warmed-up" clearance from that number.

On LS-series aluminum engines, it is common to see clearance ranges of 0.0012 to 0.0020 inch, if we add 0.0010 inch to account for engine thermal expansion, we have 0.0022 to 0.0030 inch warmed up. On an aluminum street motor, try to stay in the 0.0015- to 0.0020-inch range when building the engine. It doesn't hurt to set the no. 3 bearing to be a tad looser, because oil must lubricate the thrust bearing surfaces there as well. Keep in mind that the first set of main bearings you install provides a starting point for initial bearing clearance measurements. It is dependent on what this clearance is as to whether you need to increase clearance, decrease it, or leave it as is.

When using standard-size bearings, bearing manufacturers offer + or - 0.0010-inch bearing increments, with bearing actual clearances varying by brand as well. Sometimes mixing and matching bearings between brands and sizes is required to achieve correct clearances, although this is less likely to happen with premium, precision-made crankshafts. It is commonplace to tighten the LS-series aluminum-block main bearings up 0.0010 inch to bring oil tolerances within desired range.

When using a cast-iron block, the standard bearing clearance values apply as if you were building an LT1 or SBC. For an iron block, main bearing clearances can be in the 0.0020- to 0.0025-inch range with up to 0.003 inch allowed as leeway. Connecting-rod bearings can be treated with the same clearances. You don't want to intentionally set the clearances tight like that of the aluminum block. Bearing failure would be the likely result; 0.0020 to 0.0025 inch is generally acceptable.

The minimum cam bearing journal clearance is 0.0020 inch on a stock engine, and 0.0030 to 0.0040 inch is acceptable and the norm for a new build. As much as 0.0060 inch is allowed. Note that larger clearance tends to sacrifice a certain amount of oil pressure. Much like main bearing clearance, you may have to switch brands to gain the desired clearance. Oddly enough, the OEM GM bearings seem to be the most consistently sized, although they are much more expensive than aftermarket camshaft bearings. Remember, a new GM block, or the GMPP LSX block, comes with camshaft bearings. OEM blocks will have the camshaft bearings installed, while the GMPP LSX block will come with an uninstalled camshaft bearing kit that matches the cam bore sizes in the block. If the block is only being honed to size, the camshaft bearings need not be replaced, but if the block is to be hot tanked, acid dipped, or the cylinders are to be bored out, they do need to be replaced.

PISTONS

Pistons don't seem to get the credit they are due. The pistons in any engine have a rather large responsibility. Not only do the pistons have to be strong enough to handle massive internal explosions while not melting, they also have to absorb combustion heat and dissipate it into the cylinder walls while transferring the expanding energy of combustion into the connecting rod, which turns the crankshaft and moves your vehicle forward.

Here's the LS1 piston as we know it, nothing fancy here: flattop hypereutectic alloy, offset piston pins, short piston skirts, and low-tension rings. Surprisingly enough this piston can withstand quite a bit of abuse before expiring. With 500- to 600-rear-wheel-horsepower LS1s commonplace, we can call this "the little piston that could, but shouldn't."

The last few LS1/LS6s received coated piston skirts and a Napier second compression ring to alleviate oil ingestion issues at high engine rpm. If you look closely, this is also a full-floating piston design when mated to the proper connecting rod. The use of round wire clips is the tell-tale sign.

Stock pistons in LS1 to LS6 engines are of a hypereutectic flattop design. This is basically a cast piston with a high content of silicon, which adds more strength to the aluminum alloy than a eutectic piston (cast aluminum) would have. The stock pistons are not known for being durable in the higher power levels; they seem to die a painful death when given the right dosage of lethal power-adders. The main killer is detonation. Detonation can occur in a naturally aspirated engine with a low-quality fuel and high compression, but it more commonly occurs in forced induction cars where boost isn't kept at a reasonable level. Tuning plays a key part, but detonation can also occur with a perfect tune if the fuel quality isn't up to par with the engine needs.

FLATTOP PISTONS

A flattop piston is the closest cousin to the stock piston design as far as compression ratios are concerned. One key feature of any aftermarket LS-series piston is the addition of valve reliefs; these are a much-appreciated addition, as the stock pistons don't leave room for piston-to-valve clearance required to use the popular large camshaft profiles. The flattop pistons are commonly available in 2618 and 4032 aluminum alloys. The 4032 is ideally limited to naturally aspirated engine setups and mild power-adders, as detonation will quickly hurt these pistons if you are not careful with tuning. The 4032 is much stronger than the stock pistons and can handle power when used correctly. The 2618 forgings can be used in any setup. With the right ring gaps and proper fueling, these pistons can go well in excess of 1,000 rear wheel horsepower. Many manufacturers rate these off-the-shelf pistons for a conservative 750 horsepower, though. Valve reliefs on flattop pistons can vary between 1 cubic centimeter and 4 cubic centimeters of sizing, which will lower compression slightly. This can be reversed by cylinder head milling or slightly thinner head gaskets.

DISHED PISTONS

Dished pistons can have a variety of dish sizing. Many times dishes are called inverted domes and can be anywhere from an 8-cubic-centimeter dish to 28 cubic centimeters and more if custom-designed. The main reason for a dished piston is to lower the static compression ratio, such as when building a stroker engine that will retain a stock compression ratio or when adding a boost setup that will see street duty and pump gas use. Race-only boost setups still use lowered compression when compared to an all-out naturally aspirated engine compression ratio, but remember to factor in that many race-only boost setups use racing fuels and an extremely efficient intercooler design. Race-boost compression ratios hover in the 10 to 1 to 11 to 1 compression ratio, while street-boost applications will operate safer with 8.5 to 1 to 9.5 to 1 desired compression.

DOMED PISTONS

Domed pistons, or pop-up pistons as some would call them, are a flattop design with more material added to the top. These are measured in cubic centimeters like the dished pistons, but instead of lowering compression, they raise it. These pistons are intended for naturally aspirated or nitrous oxide setups. You can easily get another point of compression out of your engine by running an off-the-shelf domed piston. With a custom 3D domed piston that matches your combustion chamber, another 4 points of compression can be added. Depending on your engine setup, each point of compression can add 20 to 25 horsepower by itself with no other changes (with the appropriate octane fuel). You need to know which cylinder heads you are going to be using, as the dome on the piston is likely to interfere with the combustion chamber if you use the wrong part number or dome profile. Many piston manufacturers keep in-stock pistons for all common cylinder heads. Be sure to research your selection and even ask someone who knows.

Wiseco is on the leading edge of piston design. Its ability to design a piston, forge it, machine it, and then test its durability without leaving the building carries some weight. Wiseco offers LS pistons in a variety of bore and stroke sizes in the stronger 2618 aluminum alloy. This piston is the Wiseco 347-cubic-inch flattop offering: 3.905-inch bore, 3.622-inch stroke. *Wiseco*

Wiseco offers a multitude of piston designs, and if it doesn't have what you need, the company will custom-build anything. This is a 2618 forged dished piston. Dished pistons are meant to lower your compression by removing volume. They are intended to make a stroker engine live on with a more streetable compression or to build a boosted engine. *Wiseco*

This is from Wiseco's newest line of pistons called the "Flow-Dome." The dome area displaces material lost by cutting valve pockets in the piston and helps performance by improving airflow into the cylinder by matching the valve-angle profile to the piston dome. Also shown here is Wiseco's coated piston skirt, which helps alleviate engine noise annoyances such as piston slap. *Wiseco*

DESIGNING PISTONS AND PISTON MATERIALS

In talking and visiting various piston manufacturers over the course of writing this book, I learned a few new things. Pistons are more complicated than you can ever imagine, and I now know why there are only a handful of manufacturers of high-performance pistons.

The two popular piston alloys are called 2618 and 4032, each with benefits and drawbacks. The 4032 is more brittle when compared to the 2618 and has a higher silicon content, in the 11 to 12 percent range. The silicon content close to the piston surface creates a harder surface with increased lubricity. This is great for naturally aspirated and mild power-adders. The 4032 has a lower expansion rate, which allows tighter piston-to-wall clearance. It also has lower wear properties, resists scuffing, and is less noisy in use than the 2618. This piston alloy is completely fine in the right hands with good tuning and in a mild setup. Because of its high silicon content, cracks develop easier in 4032. Once a crack develops in 4032, it shoots from one silicon particle to the next in a violent connect-the-dots game that results in a shattered piston and cylinder wall damage in extreme cases.

The 2618 piston is generally the piston alloy a step above the 4032 for strength. The 2618 is 16 percent stronger than the 4032 alloy, although it's 3 percent heavier by volume. The 2618 will not shatter, but it will crack.

The crack will migrate from the high pressure area to a lower stress area and then stop, usually preventing major damage other than to itself. The 2618 is the alloy of choice for pistons in boosted and nitrous applications. Modern piston skirt coatings add lubricity and surface hardness that compensates for the lack of silicon in the piston itself.

Most LS-series engines with less than 4.100-inch bore size pistons will have a top land approximately 0.050 inch smaller than the base of the piston skirt to allow for expansion caused by the direct heat of combustion. Combustion heat is dissipated through the rings, lands, piston skirt, and engine oil. Because heat is wicked away progressively down the piston, the piston can be progressively larger toward the bottom to reduce cold running piston clearances. This taper is evident when comparing the side profile of the piston to a straightedge. Piston manufacturers tune the taper of the piston skirt for the different expansion rates of 4032 and 2618 alloys.

Also, because of the structure of the pin bosses within the forging, manufacturers add material density at the sides. To compensate for this, the piston skirt must be slightly oval (narrower at the sides) when cold to achieve a perfectly round shape when at normal operating temperature. All portions of the piston run at near zero clearance when up to temperature, if designed properly.

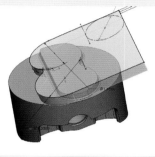

Wiseco uses CAD and FEA software to design a piston that can be tested in a virtual environment before making it to production. Here the graphic representation of the programming routine for cutting the piston notches is shown. *Wiseco*

This is the finished CAD product before the finished piston product is made. Notice this is the new Flow-Dome piston from Wiseco. This profile promotes increased flow by blocking off the dead area directly under the valve face when open. *Wiseco*

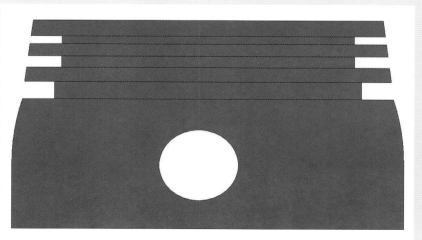

This exaggerated side view of a typical piston shows that the piston crown is the smallest diameter area by design, allowing thermal expansion caused by combustion in the cylinder. As the piston reaches operating temperature, this taper expands to a width similar to the rest of the piston. *Wiseco*

Many people think that they have a collapsed piston skirt the first time they move a forged piston around in a bore when the engine is cold; however, this is a normal byproduct of a properly designed piston. The crown of the piston has a smaller diameter than the skirt and will exhibit "piston rock" when moved from its minor thrust surface to its major thrust surface by hand (the major thrust side is the side of the cylinder that the piston is forced toward during combustion when it changes direction and is levering against the crankshaft). To check piston-to-wall clearance properly, a micrometer and dial-bore gauge must be used. Measure the piston in its largest skirt width 90 degrees from the piston pin orientation, depending on the specific manufacturer-recommended location.

Many aftermarket pistons use an offset wristpin in street duty applications. Because the crankshaft rod pin, when not at bottom dead center or top dead center, is always at an angle compared to piston travel, the piston is forced opposite of the rod pin location, using the connecting rod as the line of force. When the piston travels up, this force is directed against the minor thrust side of the cylinder. When the cylinder ignites and the piston is traveling down, it is forced against the major thrust wall of the cylinder.

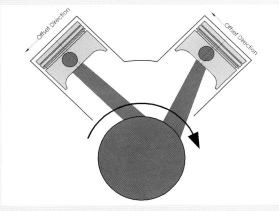

This simple exaggerated diagram shows an offset piston pin. The connecting rod is not being pushed down straight; this forces the piston against the cylinder wall at an angle. With the rod not being perpendicular to the piston, the piston is forced to the major thrust side. The offset pin helps to cure this condition. *Wiseco*

The main reason for offsetting the piston wrist pin is to reduce the noise of piston slap. Without the offset wrist pin location, the piston will slap the major thrust side of the cylinder wall when transferring its pressure from the minor thrust side to the major thrust side at TDC.

The way the offset wrist pin design works is quite simple in function but difficult to understand. As the piston changes direction at TDC, the cylinder fires, and combustion pressure pushes down on the piston, that force rocks the skirt of the piston to the offset side to come into contact with the major thrust side of the cylinder wall first. As the piston skirt makes contact, the skirt profile gently rolls the piston into further cylinder contact as the piston retreats. If the piston rolls into place rather than being abruptly forced into place, piston slap noises are reduced significantly.

The typical offset for pistons is 0.030 inch toward the major thrust side of the cylinder. If the car is strictly race-only and makes more than 1,500 horsepower, this amount of offset may be reduced.

PISTON PINS: DESIGN, STRENGTHS, DIMENSIONS, AND WEIGHT

Piston wrist pins are an often overlooked area of the engine buildup that should be given more attention during the selection process. Luckily, there are only a few choices to simplify the decision, unlike most other engine parts.

The main differences between wrist pins are the material, length, and diameter (although the dimensions that come into play are dependent on actual piston design and manufacturer). Most standard aftermarket LS-series wrist pins will be 0.927 inch in diameter and either 2.500 inches or 2.250 inches in length. The weight of the wrist pin varies with the dimensions and wall thickness. The factory wrist pin is a beefy 0.945 inch in diameter. At the other end of the scale, it is common on high-rpm naturally aspirated engines to use a 0.866-inch wrist pin for weight savings.

The shorter the length of the piston pin, the less it will bend under load. Keeping the wrist pin from bending helps prevent the piston itself from bending and cracking. If the shorter pin has enough material in the bearing load area, the shorter-length pin is always better for strength.

The standard pins for most "shelf" pistons are the 5115 and 8620 pin materials. Upgrading to a thicker 0.200-inch wall, case-hardened, tool steel pin such as the 9310 alloy is recommended for heavy shots of nitrous or large boost setups. Many engine builders will recommend the flexible case-hardened pin over the through-hardened pin. The through-hardened pin is more brittle than the case-hardened pin and will not flex. The standard 8620 wrist pin from Diamond and the 5115 wrist pin from Wiseco have both withstood more than 1,100 boosted wheel horsepower without failure. This feedback shows the conservative overbuilding and underrating tendencies of aftermarket manufacturers.

An added bonus to the shorter 2.250-inch wrist pin is that cylinder no. 8 piston in a stroker motor does not need modification. When using the 2.500-inch length pins and extra stroke, the no. 8 piston boss tower needs a notch cut to clear the LS-series crankshaft-mounted reluctor wheel. Wiseco designed its LS strutted piston forging around the 0.927x2.250-inch length, which narrowed the pin boss just enough to clear the reluctor wheel and to alleviate additional machining.

PISTON RINGS

Piston ring technology is always moving forward and improving. There are a few choices and newer technology that can be used, making the choice of rings for your engine a little harder. Desired ring types can easily vary between a naturally aspirated engine setup and power-adder setups. You want enough ring strength to more than withstand the abuse of demanding conditions.

Plasma moly top rings are a safe and basic choice that will seat quickly and last a long time, even with less than ideal honing machining. There is a little range-of-error for cylinder wall taper and out-of-round. Plasma moly tends to absorb the imperfections without hassle. These rings are an economical performance ring set that will withstand naturally aspirated and mild power-adder setups. When coupled with a Napier second ring, this ring package is a great, economical setup for engines matching its capacity. Plasma-Moly would be a great ring choice for engines producing under 600 wheel horsepower with no detonation. Over that level, an upgrade is recommended.

Nitrided-steel top rings are the best all-around top ring. They withstand huge amounts of abuse and break-in well with proper machining techniques. They are as at home in a 400-horsepower engine as a 1,500-plus-horsepower engine. Compared to plasma moly, the steel top rings will last longer given the same conditions. Because these steel rings are made of a harder material, the bore needs to be exactly right when machined or the rings will not wear-in properly. With steel rings, the cylinders and rings need to be very round, and there needs to be a minimum of 90 percent "light tight" of cylinder wall contact from the ring before break-in.

"Light tight" is piston ring terminology for showing how true the two concave ring and cylinder surfaces meet and seal. Due to machining tolerances, these are not exactly perfect until the engine is broken in, although the better these surfaces seal when new, the quicker the engine will be broken in. "Light" is where the ring is not completely against the cylinder wall. "Tight" is the surface that *is* contacting the cylinder wall. Torque plate honing is a necessity when using very round steel top rings. Shine a flashlight underneath the cylinder to check light tightness. When a piston ring is placed in the bore with a torque plate installed, we don't want to see much light between the ring sealing surface and the cylinder. Top rings are slowly getting thinner and thinner as technology proceeds: 1.5 millimeters was once the standard ring size, 1.2 millimeters is the new standard, and 0.8 millimeter may arrive soon for the street guys, although it is common to use such thin rings in race applications.

Second-ring technology has also progressed. For many years the second compression ring was just that, a compression ring. The second ring does a good job of scraping excess oil off the cylinder wall when functioning correctly. The taper-faced second ring has proven adequate for the most part, although because LS-series engines often have oil consumption issues when modified, the upgrade to a ductile iron Napier-style ring is a huge benefit. This effectively makes the second ring a dual-purpose ring as it still seals compression to a minor extent, but also acts as an improved oil scraper for what gets by the actual oil rings.

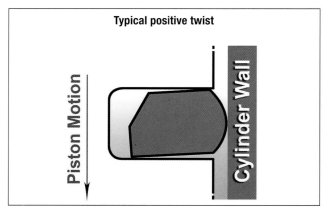

This is an illustration of the top piston ring. The positive twist compression ring seals in three points along the ring surface: the top of the ring land, the bottom of the ring land, and the cylinder wall itself. The top ring's primary function is to stop combustion from getting past the rings. If the second ring and oil scraper rings fail, the top ring will not stop much oil from entering the combustion chamber. *Wiseco*

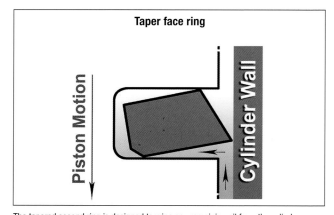

The tapered second ring is designed to wipe any remaining oil from the cylinder wall that sneaks by the oil scraper rings. The second ring, while being known as a compression ring, has evolved into more of an oil ring than a compression ring. *Wiseco*

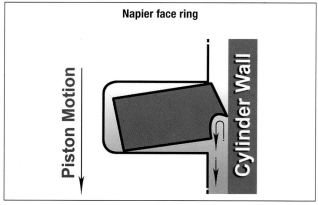

The Napier design second ring is an extension of the tapered second ring. It takes things a step further and promotes better oil sealing and reduced oil consumption than typical second rings. GM has been using this design ring in certain vehicles to promote less oil consumption. *Wiseco*

One key point to remember is the ring needs to seal against the cylinder wall and against the piston ring land at the same time. If the second ring scrapes oil off the cylinder wall and diverts it under the ring into the ring land, and if the ring does not seal in this location also, the oil will take the scenic route behind the rings and consume oil. This is also a good reason to be sure the side clearance on the rings is correct. The face of the ring against the cylinder can seal close to 99 percent, but if the side clearance is incorrect or improper rings are used, the ring itself will become an oil pump, pumping oil behind the ring and then past it when the ring shifts when the piston changes direction.

The third ring set is made up of the scraper oil rings, which are designed specifically to wipe oil off the cylinder walls. The expander ring forces the scraper rings against the cylinder wall. These, too, benefit directly from perfect machining practices.

The more round the cylinder, the less the oil expander pressure needs to be. Many people don't accept new technology and still request ring sets that worked on their old SBC or BBC. The current "standard-tension" 3-millimeter oil rings have less tension than the old 3/16-inch "low-tension" rings because of greater control in machining processes. Now we have 0.070-inch thinner standard rings that are more efficient at doing their job than the best oil rings from a decade ago.

One guideline to remember is that no two piston ring gaps should be lined up vertically to one another. They should all be staggered, with the gaps of the two compression rings 180 degrees apart and the two oil ring scrapers at least 90 degrees apart. When two similar rings line up, either oil gets into the combustion chamber, or blowby travels into the crankcase. Neither is an ideal nor recommended situation for a fresh engine.

The crankshaft with weights attached is spun on a balancer, which calculates by vibration where weight needs to be added or removed. Notice the lack of a crankshaft pulley/balancer and the flywheel. These are neutral balanced and do not help nor hurt the engine balancing, so they are left off.

Most crankshafts will need to have weight removed from the crankshaft counterweights to offset the weight of the rod and piston assembly. When drilling is necessary, you are removing material in a set point to reduce the weight in that location. Shop rags are used to protect the crankshaft journals from drill debris.

Chapter 3
Cylinder Heads

Now that the plan for your new LSX-based engine is starting to come together, you should have a good idea of which direction you are heading in. (It's best to have a plan of attack before buying *any* parts.) At this point, you may already have an idea of which cylinder heads you will ultimately be using, but keep an open mind because the right cylinder heads, along with the proper camshaft, can make or break your engine's setup. As with pistons, camshafts, or any other complicated engine part, an entire book could be dedicated to the cylinder head itself. Thus, I am forced to make some generalizations, but I hope I will give you enough knowledge to guide your engine-building path.

Some major aspects of cylinder heads include:

- The ability to properly cool the engine from combustion by providing a cooling system path around the combustion area and exhaust ports
- Providing a path for the air and fuel charge to enter the cylinder itself
- Providing an exit path for that air and fuel charge after its been combusted
- Sealing that violent combustion event inside the cylinder

Some lighter but still necessary duties include housing the intake and exhaust valves and providing necessary structure for the rocker arms and valvetrain.

Five-Axis CNC machines replicate a port based on a digitized master of a ported runner. That digitized data can be used to machine the cylinder head ports in minutes versus days for hand porting.

Long before getting to this point of assembling your engine, you should sit down and calculate your needs based on the entire combination of parts. Factors such as compression ratio, power-adders, camshaft design, and budget often dictate which direction to "head."

The combustion chamber pictured here provides a home to parts under demanding, violent internal conditions. Spark plugs, valves, and head gaskets are all at the front line of battle.

Opposite the combustion chamber and deck side of the head is the valve spring and corresponding hardware. The rocker arms are housed on a rocker rail and then bolted to these threaded pedestals providing a lever system to transmit camshaft action into the valves while fighting the pressures of the valve springs. This example is an LS6 head.

The LS7 casting uses an offset intake rocker location and offset runner to clear the huge ports designed into the LS7 head. LS7s are unique as they do not use a rocker rail system. The factory rockers are bolted directly to the cylinder head. These particular heads also have upgraded bronze valve guides installed.

For good reason, it is often said that the cylinder heads and camshaft are the two most important items for building power in an engine. Where else in an engine can you pick up 50 to 100 horsepower without power-adders by just swapping one or two components? It is crucial to understand the importance of cylinder heads when building any engine, especially if you are after every last horsepower available. You obviously don't need to spend $10,000 on a short-block if that doesn't leave you enough to spend on a good, matching pair of heads. You would be better off purchasing a more economical (or smaller) short-block and spending the difference on better cylinder heads. One of the major factors in determining which type of heads you can use is first and foremost your engine budget. This, unfortunately, dictates much of your desired performance choices.

The bottom end of your engine may be a used $300 short-block you bought from your buddy, but with the right heads on this budget-type engine you can easily make the same and often more power than a forged piston engine of similar displacement. Ported heads will make or break the setup depending on how matched your combination is. The setup that a 408-cubic-inch twin-turbo Corvette has may not be exactly ideal for your Camaro's high-rpm solid-roller 347-cubic-inch screamer.

There are many performance-oriented cylinder heads, from inexpensive ported stock-casting heads to six-bolt style, thick-deck cylinder heads. All have their place in this LS-series world and will work well when matched to correct setup, including both engine and chassis. You don't want to know how many times I've seen people come in and ask for a horsepower number, and then, after they get to that number, they magically think they have a 9-second car. How

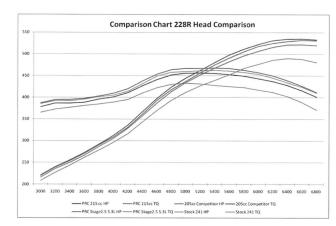

Many cylinder heads for engines of similar sizes make within 10 horsepower of each other. Shown are sample gains from stock heads on a 5.7-liter LS1 to several popular aftermarket cylinder heads, while using a small average-sized camshaft: 228°/228° .588"/.588" 114° LSA +4° advance.

disappointing it is to see 700-horsepower cars going low 12s and high 11s in the quarter, when a 400-horsepower setup could net an easy 10-second car. Match your vehicle to your new engine setup, and you won't be disappointed. I cannot stress that enough with any performance application.

In the LSX's infancy, you had little choice in which heads you could procure for your LS1-equipped vehicle, although the stock units were killer heads already. CNC-ported cylinder heads now perform as well as or better than the best hand-ported LS-series heads from just a few years ago. In fact, it is getting harder and harder to find true hand-ported heads. The efficiency and consistency of CNC machines has taken over that role.

This futuristic space-era movie prop is actually a Centroid 5-axis CNC machine. It can first digitize an existing ported head, and then that data can be changed or tweaked within CAD software before being replicated into another cylinder head casting. Once a port is digitized, minor necessary changes can be made without touching a grinder.

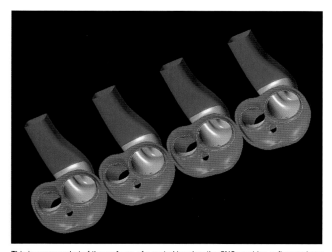

This is a screenshot of the surfaces of a ported head as the CNC machine software views it. This programming file has the intake runner, exhaust runner, and combustion chamber all tied together. With this method, an unported head is bolted in the fixture, the CNC machine set to "go," and three to four hours later, a finished, ported head is ready.

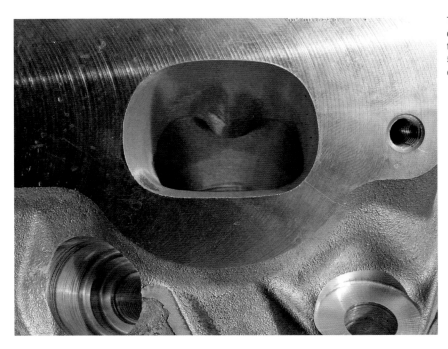

The exhaust port flange is shown here. Each LS-series cylinder head uses a different shape exhaust port than the next. The LS6 uses a D-shaped exhaust, while the 5.3-liter head shown later uses more of a rounded exhaust port.

On the intake side of the LS6 head the machining lines viewed here are left from the porting process. The lines don't affect airflow much, as there is a boundary layer of the air and fuel mixture that separates the port wall from the moving air. The spots with no machining are actually areas that are minutely larger than the intended port size and thus were not enlarged during machining.

In addition to aftermarket casting heads, a handful of cylinder head manufacturers continue to port the stock heads, such as the LS6 5.7-liter and LS2 6.0-liter 243/799 casting heads known widely as LS6 castings. Many others opt to port the 4.8-liter and 5.3-liter truck cylinder heads with stellar results. With the LS6 and LS2 and 5.3-liter castings, some "free" power can be had, as the combustion chambers on these heads are smaller than the stock LS1 heads. This adds an easy half-point of compression to your current figure.

This close-up of 5.3-liter exhaust port shows the same CNC tooling marks, but in a smoother port design.

The newest Gen IV heads designations are not left behind in these endeavors. The Gen IV rectangular-port cylinder heads will hold their own and, often before being ported, will outflow high-dollar ported cathedral heads. The Gen IV rectangle-port heads are best suited to Gen IV-sized engines with a 6.0-liter 4.000-inch and up bore. You cannot use a 6.2-liter cylinder head on a 5.7-liter-sized engine (3.905-inch bore size). It just wouldn't be ideal, and there are many other choices out there that would work much better. The only exception to this would be GM LS7 heads, as they need a 4.100-inch bore size minimum to perform as intended. While you can't effectively bring a large cylinder-bore head to a smaller-bore block, you can often bring a small-bore head to a larger-bore block with no adverse performance effects, but you are limited by the smaller valve size and runner. Ideally, you should open up the combustion chamber to match the new bore size. There are quite a few LS6 heads on 427-cubic-inch and bigger engines that perform just fine, but LS7-based cylinder heads would give the best performance on that application.

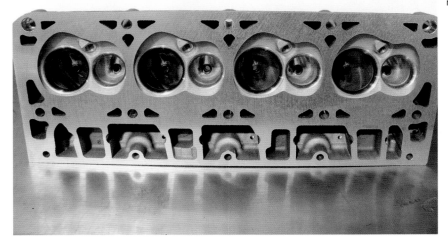

Here we see a Gen IV head known as the L92 or, more commonly, the LS3. This is already a performance-oriented head in stock form, and much like the LS7 has an offset valvetrain to accommodate the huge intake runner. The L92 heads here have a larger combustion chamber. Milling the heads may be necessary to get the desired compression ratio.

The LS3/L92 combustion chamber uses a dual quench pad shape for efficiency. This chamber is unported, and although porting the chamber reduces compression, the chamber is an extension of the valve job. Flow can be improved greatly when blended by a skilled head porter.

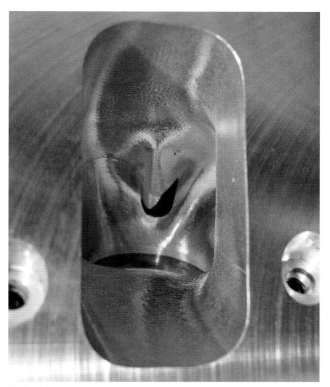

The L92/LS3 intake ports are vastly improved over cathedral runner heads. These rectangle ports are not too shabby in stock form. They are pictured here with a master port tweaked by a skilled head porter. This work can be duplicated many times over by a CNC machine.

Similar to the LS6 exhaust, the L92/LS3 borrows its D-shaped port from its older brother, sized in proportion for the larger motors now offered.

Precision Race Components designed these LS7 cylinder heads. The heads are CNC-ported like all others. (In fact, the factory versions are already CNC-ported.) PRC adds another 40 cfm at peak.

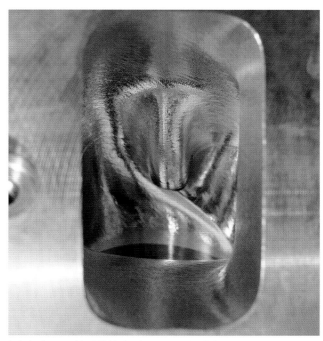

In the intake port of the PRC LS7 head, you can see the CNC porting that helps this head flow just shy of 400 cfm. The LS7 uses a dorsal fin design after the valve guide to induce swirl. With a 2.200-inch intake valve, not much gets in the way of incoming airflow.

As a rule, Gen III engines were not available from the factory with rectangular port heads, unless you include the C5R cylinder heads, which were obviously not production units. Some early Gen IV LS2 engines were factory equipped with LS6 type cathedral port heads, but the LS3, L92, L98, and LS9 engine designations all use rectangular intake ports. GM no longer offers cathedral-port heads in performance applications, other than on certain trucks and GMPP crate engines.

There is no one best head for all engines; there are just too many variables in engine setups to say that one size fits all.

As a general rule, the bigger the engine, the more airflow the engine requires on both the intake and the exhaust. That's the easy part. Naturally aspirated, supercharged, turbocharged, and nitrous-equipped engines all have different requirements, especially pertaining to cylinder heads. While blown and turbocharged engines may have largely identical bottom-end short-block components and compression ratios, they may need different cylinder heads (and camshafts) to make power efficiently. The same ideals apply to naturally aspirated vs. nitrous applications.

Here's a comparison of the stock two-piece valves (rear) to much stronger stainless-steel one-piece forgings from Ferrea. Beyond upgrades to materials and valve sizing, two-piece valves have the stem and face of the valve welded together and under demanding conditions can break apart.

The Ferrea valve has an undercut stem. This helps reduce valve stem area in the intake when the valve is opening. Looking farther, the swirl pattern under the head helps induce flow, and the 30-degree back cut gives an easy airflow gain at mid-lift points.

The smaller exhaust valve is where using a better material comes into play. The exhaust gets hotter the more power you make, so the better the exhaust valve material, the longer the valve (and sometimes the engine) will last. With exhaust heat in nitrous cars approaching 1,800 degrees F, you definitely want to make sure the valve is up to the task.

From left to right are the stock LS1 spring, "blue stripe" 918 and the newest 918 beehive springs. The beehive design helps to lighten the valvetrain mass by employing smaller diameter coils at the top of the spring and requiring a smaller retainer.

Pictured left to right, the double springs in the rear are AFR, PRC Gold, PRC Platinum, and TrickFlow. The beehive springs are "blue stripe" Comp Cams 918, and the upgraded and polished Comp Cams 918 is on the right.

This is a typical set of LS-series valve spring components. The double 1.300-inch spring is at center, the hardened lower spring seat and locator are to the right, and a lightweight titanium valve spring retainer is to the left of the spring. Keepers hold everything together on the cylinder head.

CYLINDER HEAD—INTAKE RUNNER

We'll start off with the most basic engine to try to explain what happens with cylinder heads, starting off with the intake side. The popular 5.7-liter LS1 is 5,700cc in displacement, which means that each cylinder displaces 712.5cc in swept volume from TDC to bottom dead center (BDC). Given that the stock intake port on an LS1 cylinder head is roughly 205ccs, this simply shows that the intake charge volume includes much of the intake manifold runner air (507.5cc). This means that 29 percent of the air and fuel charge entering the engine is from the cylinder head intake runner size itself, but it is important to note that 100 percent of the airflow passes through the intake runner. Much more violent airflow conditions exist during intake stroke, further complicating these numbers. These examples just show that the air needed for combustion doesn't come entirely from the cylinder head port. The same calculation performed on a bone-stock LS7 427ci, returns a percentage, 30 percent, that is quite close to the 5.7-liter 346-cubic-inch engine's ratio.

Obviously, you do not want a stock 5.7-liter-sized ports on your 7.0-liter, 427-cubic-inch engine. The intake runner volume should reflect the intended engine displacement. A larger duration camshaft mimics much the same airflow demand as a larger runner head and vice versa. Now, these runner-cubic-centimeter-to-cylinder-cubic-centimeter ratios don't mean a whole lot. I'm just trying to emphasize that there is more to a head's performance than the heads themselves. Outside factors can affect power output just as often as mismatched internal engine components.

Finding the right cylinder head and camshaft is a balancing act. The larger the runner, the slower the velocity (air speed) is. The trick in cylinder head design is to maximize airflow, while minimizing runner size to maintain velocity.

CYLINDER HEAD—EXHAUST RUNNER

The exhaust runner is always the smaller of the two runners in a V-8 designation engine. Much of the reason it is smaller is due to the same factors as why a combustion engine pushes the piston down when it fires. That combustion is still pressurized when the piston is approaching BDC while the exhaust valve is opening. The burnt combustion is pressurized, and the piston is forcing the remaining gasses out through the exhaust valve. On the exhaust side, it's a fairly simple process: exhaust valve opens, piston pushes exhaust out. Only if it were all that easy....

Next, the intake valve is opening while the exhaust valve is about to close. The start of the four-stroke cycle begins following the end of the previous four-stroke cycle, which ends with the exhaust stroke. The time period when both valves are open is called *overlap*. This phase is also known for its scavenging effect. The leftover exhaust charge that is well on its way out will help pull some of the new intake charge into the cylinder. Some also inevitably gets pulled out with the exhaust. Proper camshaft timing events help to limit this

happening, as fuel in the exhaust does very little for internal engine power output.

The reason to mention overlap with cylinder heads is that there is some control over this scavenging effect with valve sizing and exhaust port sizing. Smaller valves and exhaust port design can limit the overlap effect. Remember, with a blower, airflow is pushed into the engine rather than drawn in. Well, if exhaust valve timing events are not correct—or if the exhaust valve/port flow is too great—the pressurized air in the intake tract will push air/fuel into the cylinder and sometimes right out the exhaust port! If too much fuel is present in the exhaust, it can begin to glow, much like a lean condition. But it's not lean if you're providing the fuel for this incinerator, now is it? This is a form of forced *over-scavenging*. Some is OK and normal, but excessive amounts of fuel in the exhaust port will still ignite and burn. However, it will happen in the exhaust itself rather than in the cylinder where it should be. Though on an N/A, or nitrous, engine this anomaly may occur only at low engine speeds with larger-sized camshafts with big overlap numbers, you still need to pay special attention to port sizing when picking your cylinder heads for different applications.

It is interesting to note that while many may assume that blowers and turbos operate under the same internal conditions, quite the opposite effect happens during overlap with a turbo setup. The highly pressurized exhaust from the hot side of the turbo system will attempt to flow back into the cylinder and even past the intake valve if not set up correctly. This is due to exhaust pressure being very much higher than the "pressurized" intake and cylinder. There are a lot of variables to this backwards exhaust flow: turbo sizing, exhaust header design, downpipe sizing, etc. will all play some part. So while it is possible that the port and camshaft may be the correct size to the application, a condition external to the engine may cause exhaust to "back up" right back into the engine from where it originally was burnt! Where are all those "boost is boost" guys? Now that I have made everyone with forced induction worry, we'll move on to the easy stuff!

CYLINDER HEAD—COMBUSTION CHAMBER

Combustion chamber sizing affects the compression ratio of the engine. Although the piston and bore and stroke affect the compression the most, minor changes in the chamber size and head gasket thickness can help to tweak your compression ratio to exactly what you want. Some combustion chambers are considered an extension of the valve job, with modifications to the chamber assisting cylinder head flow.

One thing that seems to get overlooked, intentionally or not, is that the combustion chamber's width should match or be smaller than the bore size. There are a few people who think otherwise, but you should not attempt to run a 4.125-inch combustion chamber with a 3.905-inch cylinder size. There would be a huge stepped transition from the combustion chamber into the cylinder, which would be detrimental to cylinder head flow.

HEAD GASKETS AND FASTENERS

Using proper head gaskets and fasteners can be overlooked when lining up your new parts. You can build the baddest short-block and top it off with the highest flowing heads, but if your head gaskets are poorly chosen or you have insufficient clamping torque on the head bolts or studs, you may have just built a one-pass wonder that will be lucky to make it cleanly down the track twice in a row. In particular, any power-adder engine needs considerable thought given to the head gasket sealing components and everything possible must be done to keep the cylinder heads sealed to the block.

The aftermarket has grown in leaps and bounds with all LS-series family engine parts; head gaskets and fasteners are no exception. For a time, there were no aftermarket head gaskets, head bolts, or head studs available. Now there are many choices. Some are easy bolt-in upgrades, while others require machining.

Concerning head gaskets, there is something for any application—from GM multi-layer steel (MLS) to Fel-Pro to Cometic. OEM graphite gaskets are OK for mild builds and for anything naturally aspirated, but for power-adders the OEM MLS head gasket should be considered as the minimum upgrade. Keep in mind that if early LS1 or 5.3-liter head castings are used and they have a casting notch in the deck surface, MLS gaskets will not seal and the OEM graphite gaskets are recommended, unless the notch is welded in solid and resurfaced flat. If the casting notch is retained, you limit your head gasket choices to graphite or OEM graphite replacement.

The bore size is also a factor. The minimum head gasket bore size is at least the size of the larger of either the cylinder or the combustion chamber. For example, if you have a 3.905-inch bore and a 4.125-inch wide combustion chamber, you need a head gasket with a 4.125-inch bore at minimum. Similarly, a 4.030-inch bore and 3.905-inch chamber would require the 4.030-inch gasket.

Quench is affected by head gasket thickness, the block's deck height, and the compression height of the piston. Head gasket thickness affects compression ratio as well as quench, if the head gasket chosen based on the desired quench raises the compression too high, then it's best to tone it down a little and run a thicker head gasket. The minimum quench dimension on a street engine is usually 0.035 inch, which does not mean 0.035-inch gasket thickness but the distance from the top of the piston to the quench pads of the cylinder head. If your pistons come out of the bore 0.010 inch and you use a 0.055-inch head gasket thickness, your quench distance is 0.045 inch. If your piston is at zero deck at TDC, then your quench is determined directly by gasket thickness; 0.040-inch gasket would net 0.040-inch quench on a zero-deck piston. Quench is one of the easier measurements to calculate.

Spending some time with a compression calculator and realistic engine goals should help determine the head gasket choice. Once you find out what gasket gets you where you need to be in terms of compression, you also have to factor in gasket strength for your goals. One of the more popular head gasket manufacturers is Cometic, which has its own line of MLS gaskets with a ton of available thicknesses and bore sizes. MLS stands for multi-layer steel, which means that there are usually three layers to the gasket that are riveted together: an upper sealing layer, a lower sealing layer, and finally the spacer in the middle that varies with gasket thickness.

While GM MLS gaskets can be installed as is on almost any sealing surface, Cometic gaskets require a strict 50 roughness average (RA) or better on both the block and heads to seal properly by itself. The roughness average needs to be checked by a machine shop. The lower the number, the smoother the finish will be. The higher the number, the rougher the finish. Cometic MLS gaskets seal best with the smoother surface. Cometic does not recommend additional sealant on the gasket surface, but experience has shown that a light coating of copper spray to both sides of the gasket will help alleviate coolant seepage and leaks to the outside of the block without hurting gasket strength.

Options for cylinder head fasteners include the OEM "1-time use" head bolts, ARP head bolts, ARP head studs, and for those who desire more clamping load, the L19 drop-in head stud is available. All of these fasteners are the same diameter as factory bolts, which is the M11x2.0 thread size and pitch. These are all fine for a cast-iron block, but an aluminum block sometimes has extra torque values above the manufacturers specifications, which will weaken the block threads leading to head gasket issues and coolant consumption as the first signs. The M11 bolt is right at the same size as the standard 7/16ths bolt/stud.

The main difference between OEM hardware and aftermarket is the torque-to-yield design of the factory components. The aftermarket bolt choices do not yield when torqueing to specifications. Torque-to-yield bolts work fine with most naturally aspirated applications and mild power-adders, although if you're out to set the world on fire with your new engine, it is probably in your best interest to spend the added coin on stronger bolt hardware throughout the engine.

To add more reliability to the head gasket and fastener system, a complete head bolt and stud retorque is highly recommended after a few days of engine operation. This is only possible when using reusable ARP hardware, not OEM torque-to-yield bolts. Let the engine cool completely, and loosen and retorque each cylinder head fastener to peak torque value one-by-one, following the same pattern as the initial torque sequence. Only tighten and retorque the bolt you are on, not any others.

For those who want to make more than 1,000 rear wheel horsepower, the standard off-the-shelf fasteners are not sufficient, but neither are OEM-type heads. If you want to make big power and not have head gasket issues, you may want to look into a 1/2-inch-diameter head stud conversion. Kurt Urban Performance (www.kurturbanperformance.com) offers a 1/2-inch head stud kit in the 8620 ARP designation that can be installed by anyone with skill at operating a Bridgeport-type machine. It involves drilling and tapping the block to 1/2-inch threads. One of the main advantages to this upgrade is that it increases the thread contact with the block. This is ideal for any aluminum-based block, and with iron blocks it is even better. The upgrade to 1/2-inch studs also requires modified heads and head gaskets to match the stud diameter. When coupled with the six-bolt block and heads, this makes for a bulletproof head gasket sealing setup.

After much personal debate, I decided to include the big three cylinder head porters and manufacturers that exist in the LS1 community. These companies' products are the most popular in the LS-series world, not because they are the only ones there, but because they each offer something unique that other manufacturers cannot or do not provide. This book is intended for the majority of LS-series owners with street, drag, and road race cars, so not much time will be spent discussing the "one-off" or high-dollar custom hand-ported heads.

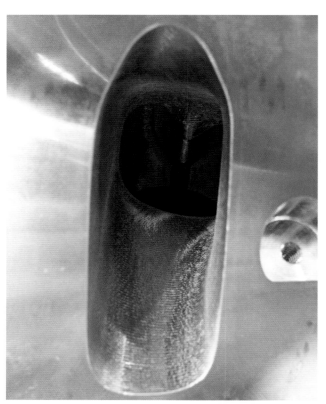

Air Flow Research

Air Flow Research (AFR) was the first company to offer an aftermarket LS-series head casting in 2004. Much time went into the development of the AFR CNC-ported "Mongoose" 205cc, 225cc, and the much-anticipated 235cc cylinder heads. These three cylinder heads can cover the majority of LS-series owner's needs. AFR's offerings for LS-series engines are based on the factory cathedral intake runner design. AFR offers a 0.750-inch-thick deck surface on its LS-series heads. AFR prides itself on keeping velocity up while still putting up really good airflow numbers. Anyone with access to a flow bench and the means to check air velocity will tell you that these heads are no slouches.

Air Flow Research stepped up to the plate with no aftermarket cylinder head competition early on. The company designed and manufactured the fully CNC-ported LS Mongoose cylinder heads in 205cc and 225cc sizes. This is the 225cc intake runner and can be used on any LS engine.

The AFR exhaust port on the 225cc heads comes in at 85cc and is also 100 percent CNC-ported. Good-sized exhaust ports are helpful in power-adder engines, more so for nitrous.

With a 4.125-inch head gasket in place, the dual quench pads are highlighted. One additional feature AFR has always promoted is its 0.750-inch-thick deck surface. The deck's added material results in substantial improvements to head gasket clamping.

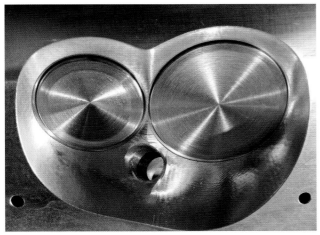

These AFR 225cc heads have 72cc chambers, and 62cc and 65cc chambers are also available. With milling, just about any size from 59cc to 72cc can be had. The 225 heads are equipped with generous 2.080-inch and 1.600-inch valves.

Much like any other cylinder head manufacturer, AFR offers a base spring package that adequately supports 0.600-inch lift, featuring dual springs, hardened lower spring seats, titanium retainers, and hardened keepers. The AFR heads are ready to bolt on and even facilitate OEM rocker arms and valve covers.

Trick Flow Specialties

Trick Flow Specialties (TFS) made its name with its popular Twisted Wedge heads for small-block Fords. Their CNC ported and as-cast LS-series offerings are never a letdown. TFS is another company that offers an aftermarket LS-series cylinder head casting and made many improvements with a clean-slate design. One of these upgrades is a 13.5-degree valve angle; 15 degrees is the stock valve angle. The 13.5-degree angle allows better piston-to-valve clearance in motors that do not have factory valve reliefs. This also allows a much larger valve to be used. TFS also improved the deck thickness, upping it to 0.600 inch for much improved head gasket sealing Trick Flow also found that they could unshroud the intake valve by relocating the spark plug. This added better midrange intake flow numbers. TFS offers a 215cc, 220cc (as-cast), 225cc, and 235cc cathedral runner heads, all for varying applications and engine bore sizes. Recently TFS announced that a six-bolt version of its popular cylinder heads would be released.

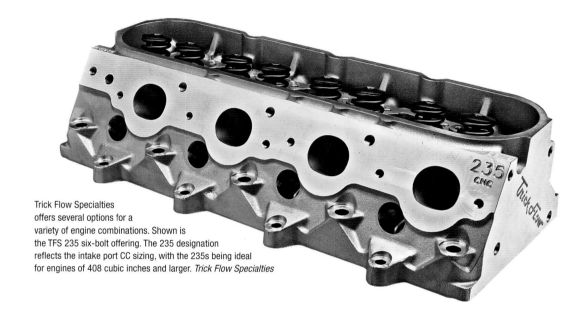

Trick Flow Specialties offers several options for a variety of engine combinations. Shown is the TFS 235 six-bolt offering. The 235 designation reflects the intake port CC sizing, with the 235s being ideal for engines of 408 cubic inches and larger. *Trick Flow Specialties*

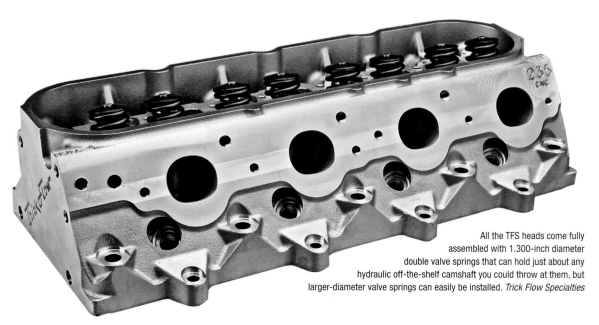

All the TFS heads come fully assembled with 1.300-inch diameter double valve springs that can hold just about any hydraulic off-the-shelf camshaft you could throw at them, but larger-diameter valve springs can easily be installed. *Trick Flow Specialties*

The TFS 235 heads come equipped with 70cc combustion chambers and can be milled smaller as needed. The valves used in the 235cc heads are 2.080-inch intake and 1.600-inch exhaust. *Trick Flow Specialties*

Viewing the exhaust port side, we can see the serious CNC port work. The TFS 235 exhaust features 80cc exhaust ports that flow some big air. We can also spot one of the extra cylinder head bolt holes in this image. When coupled with a six-bolt block, you can be assured of impressive cylinder head gasket longevity. *Trick Flow Specialties*

The intake port is where these heads shine. TFS stuck with the popular cathedral intake ports for its cylinder head offerings for greater compatibility with current aftermarket components. *Trick Flow Specialties*

Texas Speed and Precision Race Components

Texas Speed & Performance (TSP) and Precision Race Components (PRC) are sister companies that specialize in LS-series performance and LS-series performance only. In 2004, TSP and PRC, much like AFR, noted the presence of a huge customer base that sought reasonably priced, ported GM-casting cylinder heads but with higher quality standards than what was available to them at the time. PRC first offered ported 5.3-liter heads but soon expanded into LS6, LS3, and more recently LS7 castings with a few ported aftermarket castings here and there. These simple steps helped solidify their place in the LS-series world. Today, PRC CNC ports and assembles many stock-casting heads in-house on their two 5-axis CNC machines, with plans for an aftermarket casting of its own on the horizon.

Texas Speed & Performance (TSP) and Precision Race Component (PRC)'s most popular cylinder heads originate from the OEM 5.3-liter casting, which after CNC porting, bears no resemblance to its former self. With upgraded 2.020-inch and 1.570-inch stainless valves standard, these produce excellent results on any 346- to 383-cubic-inch engine.

PRC is working on an aftermarket casting of its own. Shown here is the raw casting before machining. Some highlights are a 59cc combustion chamber with no milling, additional deck thickness for power-adders, and improved spring pocket strength.

PRC's casting also features an extra-tall rocker flange for aftermarket rocker arm clearance. It is available in 215cc, 227cc, and 237cc intake port designs and with a variety of combustion chamber sizes and valve options.

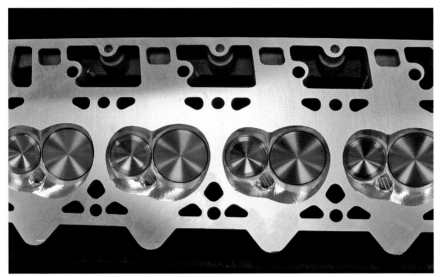

The first PRC cylinder head tested by TSP was the 215cc offering. In initial dyno tests, when mated to the TSP MS4 239/242 camshaft, more than 600 horsepower was achieved. Initial track testing also proved successful with TSP and PRC running 9.82 at 134 miles per hour with an OEM stock LS1 short-block in its test bed 1998 Camaro.

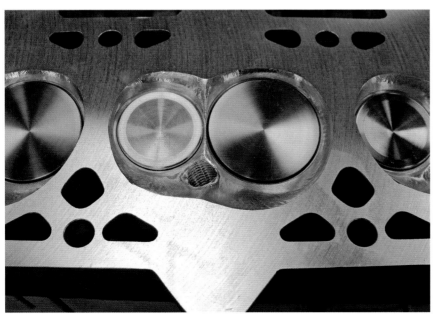

The PRC head's small-chamber 59cc design allows for increased compression using generously sized camshafts while not requiring piston fly cutting. The company is also testing a 13-degree valve angle to further improve valve clearance.

Chapter 4
Camshafts and Hardware

The camshaft is the brain of any engine and controls all aspects of the air and fuel mixture entering the cylinder at the correct moment and when the spent combustion exits the cylinder. The basic camshaft design in a production car is designed to deliver efficiency and clean emissions. Choosing a camshaft should be a direct reflection of how you want your engine to operate.

The proper camshaft should be one of the last major components you choose when planning your new engine setup. You should know engine sizing, cylinder heads, compression ratio, and intake and exhaust before fully committing to a certain design camshaft. The camshaft plays a large part in making everything work together in harmony.

Competition Cams offers numerous camshaft options, from mild to extreme, including hydraulic and solid roller camshaft grinds. They have camshaft options for the majority of LS-series engine users.

The LS-series camshaft is based on a billet 8620 core, as roller camshafts do not experience as much friction as a flat-tappet camshaft. Because the LS-series engine design has no distributor, no distributor gear is found on the camshaft.

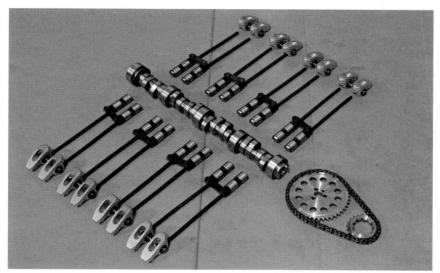

The LS-series uses a basic cam-in-block pushrod setup. This is a tried-and-true design that GM doesn't seem to want to veer from in the near future. It is a simple design that works well.

The Gen III camshaft uses a 2x camshaft reluctor gear to determine which cylinder to start the firing order. This gear is located at the rear of the camshaft on all Gen III engines. Gen IV engines use a camshaft gear-mounted reluctor wheel. The Gen III camshaft can be used in either generation of engine, but a Gen IV cam can only be used with a matching Gen IV timing set.

INTAKE AND EXHAUST DURATION

The length of time the intake and exhaust lobes keep the valves open is called duration. This time frame is measured in crankshaft degrees. There are a few ways to compare camshaft duration, but the industry standard is to measure it at 0.050 inch of cam lobe lift (not valve lift). The 0.050-inch figure is also a good reference number for degreeing the camshaft. Many manufacturers will also give 0.006-inch opening and closing duration figures, as these tend to show the actual characteristics of the camshaft design in more detail. Remember that these figures reflect cam lobe duration. To determine what the valve is seeing, you have to multiply these by the 1.7 stock rocker ratio if you are measuring duration or degreeing the camshaft from the valve side of the rocker arm. Duration at 0.050-inch lobe lift is the same as duration at 0.085-inch valve lift with a 1.7 rocker. Both are methods of measuring the duration at the camshaft but use a different measurement location.

LOBE LIFT AND VALVE LIFT

Many people focus on peak lift to judge power output or how much camshaft an engine has, but peak lift has little to do with how aggressive a camshaft is. Duration has much more of an effect on power output and aggressiveness. Valve lift is the measure of how far the valve travels from completely closed to fully open. Lobe lift is a measurement of the actual amount that the camshaft pushes the lifter. Lobe lift multiplied by rocker ratio = valve lift, and valve lift divided by rocker ratio = lobe lift. You can run more lift than exhaust lift on the intake side and improve power output to an extent on most LS-series engines. With many popular LS-series cams, valve lift is in the 0.550-inch to 0.600-inch range.

With Competition Cams, you pick out a duration, and the company gives you a lift amount proportional to the duration for which that lobe profile is designed. The CC 224 XER lobe has a 0.581-inch lift, while the CC 228 XER lobe has a 0.588-inch lift. The longer the duration, the larger the actual lobe lift. Camshaft timing is much more critical than valve lift. The other thing to keep in mind with valve lift is that if a 0.650-inch lobe makes the same power as a 0.600-inch lobe, the 0.600-inch lobe would be the better choice as it would be less harsh to valvetrain hardware.

LOBE SEPARATION ANGLE AND OVERLAP

The lobe separation angle (LSA) is the amount of camshaft degrees between the centerline of the intake lobe and the centerline of the exhaust lobe on each cylinder. I prefer to think of it as the exhaust-to-intake distance, as you are comparing the closest two actuation points on the camshaft, when the exhaust is closing while the intake is opening. Although the camshaft LSA is measured in camshaft degrees, you can convert to crankshaft degrees by simply doubling the number. The LSA, along with camshaft duration, has a direct effect on overlap, which is the duration when both the exhaust and intake valves are open.

The LSA on fuel-injected LS-series engines is usually best in the 110 to 114 LSA range for naturally aspirated and N20 engines, while turbo and blower cams usually are best in the 114 to 116 LSA range, although that alone isn't the only deciding factor. An important thing to remember is that duration plays a huge role in determining your overlap, in addition to the LSA. The larger your LSA numerically, the shorter your overlap period is. The lower the LSA, the larger your overlap is, given no other changes. Determining ideal overlap is a black art in itself.

```
PART # 54-000-11      SN#: N 8343-08
ENGINE: LS1  CHEVY LS1/GEN III '97-UP ROLLER        PART#: 54-000-11
   TEXAS SPEED & PERFORMANCE LTD
GRIND#: LS1 3734R /3288R   HR112.0
SPC INSTR 1:
SPC INSTR 2:
                    INTAKE   EXHAUST
VALVE ADJUSTMENT    HYD       HYD
GROSS VALVE LIFT    .615      .622
DURATION @
 .006 TAPPET LIFT   297       307
VALVE TIMING    OPEN          CLOSE
 @ .050   INT:    16  BTDC    52  ABDC
          EXH:    63  BBDC    11  ATDC
THESE SPECS ARE FOR CAM INSTALLED
 @ 108.0  INTAKE CENTER LINE
                 INTAKE   EXHAUST
DURATION @ .050    248      254                  SPRINGS REQUIRED
LOBE LIFT          .3620    .3660        VALVE SPRING SPECS FURNISHED
LOBE SEPARATION  112.0                        WITH SPRINGS
```

Here is a sample camshaft specification card. This card lists all pertinent information related to the camshaft design: open duration, lobe and valve lift, lobe separation angle, and intake centerline angle. The opening and closing degree readings are shown for 0.050-inch lobe lift measurements.

INTAKE CENTERLINE AND ADVANCE

The intake centerline (ICL) is often confused with the LSA. You can see where the confusion comes from because the numbers are relatively close (depending on camshaft advance). While the LSA is measured in camshaft degrees, the ICL is measured in crankshaft degrees. The ICL is a measurement of where the absolute centerline of the intake lift point is in relation to top dead center. With zero advance, the ICL will exactly match LSA, because ICL, being measured in camshaft degrees, will be half of the LSA's crankshaft degrees. If you advance the camshaft 4 degrees, the actual ICL will be moved by that same amount of 4 degrees advanced. Changing ICL or advancing or retarding the camshaft moves both the intake and exhaust valve centerlines the same amount either direction, as they will be always at the same fixed LSA spread. The LSA is permanent and unadjustable once the camshaft is ground, while ICL can be manipulated afterwards. ICL and advance are essentially the same thing in regards to how the camshaft events are affected. You can change camshaft advance (and in-turn intake centerline) by an outside means such as an adjustable timing chain set or offset cam dowel bushings.

As one cylinder fires in an LS-series engine, the cylinder that fired just before that is halfway through its power stroke. The next cylinder in the firing order is halfway through its compression stroke and will ignite in a one-quarter crankshaft rotation. This is a constant process in a running engine. In any V-8, a cylinder fires every one-quarter crankshaft rotation, or 90 degrees.

Understanding valve events helps you choose a matching camshaft to your engine application. You don't want a huge naturally aspirated "set-on-kill" camshaft behind a blower or turbocharged setup, nor would you normally want a solid-roller setup in something your wife drives to take the kids to school. There are many viewpoints and opinions on camshaft sizing, and as with cylinder heads there is no "one-size-fits-all" with camshafts. Typically, you want the camshaft with the shortest duration that provides the power level you require. It is better to go slightly too small than too big.

LS-series engines respond extremely well to all camshaft choices, provided the camshaft properly complements the rest of your setup. In a race-only setup where you can work with gearing, clutch adjustments, and torque converter sizes, it is much easier to run a larger-sized camshaft than in a vehicle that is primarily used for commuting and recreational driving.

The camshaft duration and lobe separation angle have a direct effect on the rpm band and low-rpm idling and driving characteristics. Larger camshafts are most detrimental to idle quality, as it is more difficult to fill the cylinder with a good air and fuel mixture. With larger camshafts, you have to remember that at low engine speeds, the engine is not operating in its ideal power band. Typically, any upgraded camshaft will have a hint of cam lope to it, although there are some smallish camshafts that are designed to be a bit stealthy. With creative camshaft event timing you can hide the camshaft sound to an extent, and if it is stealthy enough, it may be possible to pass your local sniffer tests.

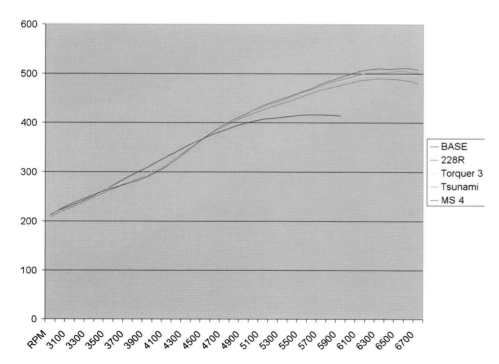

This dyno graph compilation shows how well LS-series engines respond to camshaft-only swaps. A stock long-block 5.7-liter LS1 equipped with a FAST intake and Kook's headers is the baseline. Once baseline numbers are logged, three various camshaft grinds are tested back-to-back with no other changes. These camshafts range from an average medium-sized camshaft through medium-large camshaft designations for a 346ci-sized engine.

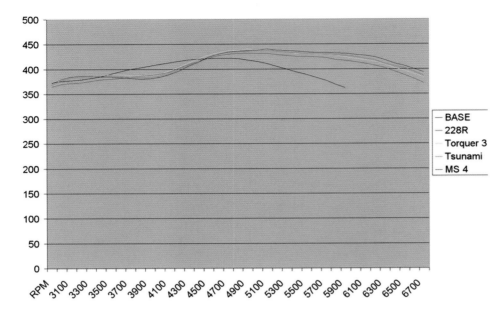

An extension to the previous camshaft dyno graph. This graph represents larger camshafts in the spectrum, designed and intended for all-out 346ci LS1 engines. Notice there is very little difference between the Tsunami 235/240 and the CC 240/244. When camshafts get to this point there is very little to gain by going bigger. The 240/244 is limited by head flow and cubic inches, which is demonstrated by the much smaller camshafts outperforming it.

The more cubic inches you have, the more camshaft your engine needs to make good power numbers. What may be viewed as a large camshaft for a stock engine may be a small or average-size camshaft for your new 408-cubic-inch stroker. Camshafts, much like cylinder heads, can be proportionally sized for cubic inches. Spend some time thinking over your goals for your new engine or engine revamp. The factors to consider are intended rpm range, power-adder or nitrous use, and finally the engine's intended use. Of course, for blown or turbo setups, you don't want to get too rowdy.

Compromises usually have to be made, as there is no one cam design that does everything you may want. While you may not be able to realize all of your goals with one camshaft, you can usually get close enough to be tolerable. Think of what is important to you in the end. Max power usually means lots of engine rpm; if the camshaft works well at high rpm, then it usually is temperamental at lower engine speeds, and vice versa. Camshafts that have good low-end usually fizzle out at high engine speeds.

LIFTERS

Another important component that complements the camshaft is the roller lifters. There are hydraulic and solid roller lifters, and each kind has to be mated with a compatible camshaft. Hydraulic lifters are designed to take up the slack between the rocker arm and valve tip. When running, they provide just enough tension to eliminate lash (air gaps) in between any valvetrain parts. Solid lifters are designed to have a small measured gap; there is no preloading of the valvetrain when using solid lifters. Solid lifters are just that—solid—while hydraulic lifters have a plunger that floats, using engine oil pressure to keep a load on things while compensating for engine wear. Solid lifters have the advantage of stability at high engine speeds. When the camshaft "requests" lift, the solid lifter delivers it directly to the pushrod and in turn directly to the valve. Hydraulic lifters operate more like a shock absorber when running. While this dampens harsh forces, this also can limit their ability to properly follow camshaft profiles at high engine rpm. The theory is that hydraulic lifters absorb a certain amount of lift generated by the camshaft. The lift that the valve sees is less than what the lift mathematically equates to.

OEM hydraulic roller lifters seem to be durable enough for performance applications, but they do have a small failure rate when used in extreme engine designations, like any lifter brand would. Competition Cams offers a direct replacement OEM lifter kit as well as a high-rpm lifter kit that drops right in. Many companies also offer tie-bar lifters that eliminate the plastic lifter trays. Tie-bar lifters are available in both solid roller and hydraulic versions.

One thing to watch with aftermarket lifters is the preload settings. There is a small window of adjustment in these, and they do not operate well if veered too far from the ideal preload settings. Because LS-series engines have a non-adjustable valvetrain, to adjust preload you either need different length pushrods or rocker arm shims and sometimes a combination of both to be perfect (if you need one of those in-between sizes). With the current availability of pushrods in 0.025-inch length increments, shim use should be rare, but it may be required with non-OEM valve lengths or different heads.

LS-series engines use a composite guide system to hold the roller lifters in line to the camshaft lobes. This method works well, and a welcome trait is that these guides can be made to hold the lifters up and away from the camshaft lobes, allowing quick and easy camshaft swaps.

With the GMPP LSX block's six-bolt heads, the conventional LS-series lifter tray design would not fit. Lifter trays for each cylinder were the inexpensive answer.

Tie-bar lifters are a more reliable design than the OEM lifter tray setup, but they are not needed for all applications. Tie-bar lifters do not require the composite hardware setup, but they also do not allow for the quick and easy camshaft swaps that we've grown to love on OEM LS-series engines.

The makeup of an OEM hydraulic lifter is shown. The lifter uses engine oil pressure to take up the slack in the valvetrain, alleviating the need to adjust the valvetrain periodically for lash. Proper preload setup is required with all hydraulic lifters using non-adjustable rocker arms such as the OEM and many aftermarket rockers.

VALVE SPRINGS

One thing you may notice is that LS-series camshafts are much bigger than conventional small-block Chevrolet camshafts. This causes much more stress on the valvetrain. To complicate it further, LS-series engines use 1.7 ratio rockers stock. The LS7 engines use an even more aggressive 1.8 rocker ratio, which adds more stress to the valvetrain. To counteract the high lift and high rpm of these engines, a beefy valve spring is needed. The stock valve springs have a beehive design, which reduces moving weight at the top of the valve, as the top of the valve and valve spring coils and retainer move much more than the bottom coils during operation. Reducing the diameter of the spring and retainer saves weight, and any weight reduction improves valvetrain operation.

Beehive valve spring technology has progressed greatly in the last few years, and beehive aftermarket springs are available from most aftermarket manufacturers. The advantages of a beehive are light weight, stable valvetrain, and somewhat economical pricing. The best LS-series beehives are rated to 0.625-inch lift. Many beehive springs are manufactured using an ovate wire compound that increases strength and lift capacity. The only problem with using single beehive springs is that if you break a spring, you almost always do damage to the valve or piston below it. Although valve spring failure is rare, it does happen randomly to any kind of valve spring. You can do a few things to prevent it, such as warming up your engine before racing and not running more lift than recommended. If a spring is destined to break, however, it will.

The use of lightweight hollow-stem valves and titanium retainers along with beehive springs makes for a high rpm-happy valvetrain, and if set up right, the lightweight single-spring valvetrain can be stable at higher rpm than a comparable dual-spring setup.

Because LS-series engines like more camshaft than most engines, people try to throw just about anything at it to find the limits. You can run a 0.650-inch lift with some manufacturers' hydraulic camshafts, which has necessitated the development of the 1.300-inch-diameter dual-valve spring kit. Crane Cams, Competition Cams, Manley, and Precision Race Components all have a version of the dual-valve spring kit for LS-series engines. All have varying specifications and hardware to complement the spring kit. Some dual-valve spring kits are rated for 0.600-inch lift, while other aggressive setups are rated for a 0.650- to 0.660-inch maximum lift. To run these crazy lift numbers, the rest of the valvetrain needs be evaluated carefully. Pushrods, valves, lifters and preload, and rocker arms all need to be exactly spot-on to run these high-lift camshafts. Pay special attention to spring rate, spring installed height, and coil bind numbers.

Special care needs to be applied with solid roller camshafts when setting up springs and valvetrain. Solid roller cams require lots of spring pressure for the lifter to follow the aggressive designs of these camshafts. Sometimes longer valves are needed along with really stout rocker arms, and usually shaft-mounted rockers are needed for uniform strength throughout the valvetrain.

All Gen III/IV LS-series engines use modern beehive-shaped valve springs. The main advantage to using beehive springs is reduced valvetrain weight because of the smaller top spring coils and spring retainer. The yellow spring is the LS6 valve spring that was required for the C5 Z06's and LS2's larger-lift camshafts.

Beehive valve spring technology has also progressed to the aftermarket. Competition Cams saw the potential of the beehive design and focused on improving it for the race-oriented crowd. The old Comp 26918 Blue Stripe on the left is rated for a 0.600-inch lift, while the upgraded 26918 CC spring on the right is rated for a 0.625-inch lift.

When the single beehive spring doesn't cut it, an upgrade to one of the popular dual-spring kits is the next step. The Precision Race Components Gold and Platinum spring kits deliver long life, improved spring pressure, and the capacity for a 0.660-inch valve lift when coupled to a matching valvetrain setup.

ROCKER ARMS

GM's engineering crew really had their cards in line when they designed the stock LS-series rocker arms. The OEM rocker arms are a 1.7 ratio on everything except the LS7 (1.8 ratio) and have a roller bearing fulcrum

that does well with stock and mild camshafts. While they have their weaknesses, such as spitting out rocker arm bearings, they don't have many other problems. Even the rocker arm fulcrum bearing isn't really a problem anymore. A few companies offer an upgrade to the stock

The OEM rocker arms are strong enough to handle most anything you can toss at them. The only durability problem is that the fulcrum needle bearings are always trying to escape, but even that problem has a cure.

The Gen IV LS3 and LS7 cylinder heads use relocated intake valves (left). An offset rocker arm must be used on the intake side. This side-loading accelerates the fulcrum needle-bearing escape problem. It's popular when using these heads to upgrade the rocker arms with the Harland Sharp bearing rebuild, which positively locks the roller bearings in place.

rocker with a fully enclosed bearing system, virtually eliminating all rocker problems. One thing to remember is the OEM rocker was never designed to run much more than 0.550-inch lift, so when you exceed that lift point, the non-roller-tip scrubs the valve tip much more than originally intended. This causes some valvetrain noise and eventually may wear out aftermarket valve guides due to lateral stress on the valves, although the stock powdered metal valve guides never seem to have problems with wear when using high-lift camshafts.

The LS7 rocker arm is especially vulnerable to fulcrum bearing failure because of its aggressive offset. 1.8 ratio versus the 1.7 ratio on all other LS-series engines. A stock LS1 rocker is at left, while a similar LS7 Harland Sharp–upgraded rocker is on the right. Notice the snap rings that retain the bearing assembly, while the factory rockers use a press fit.

Comparing the OEM rocker fulcrum to the upgraded Harland Sharp bearing kit, we can see the main advantages easily. The HS kit uses an 8620 fulcrum versus the factory cast piece. The HS also uses a needle bearing and snap-ring kit that is similar to the equipment in the company's aluminum rocker arm kits.

While it is popular to replace rockers on SBC, SBF, and many other engines as one of the first mods to your engine, this has never caught on with the LS-series engines. Some of this is because of problems a few have encountered with aftermarket rocker arms, and some is because of the advanced design of the stock rocker arm system. While aftermarket rockers may have a bad rap on LS-series engines, they do have their place

when set up correctly. With proper valve spring tensions, pushrod dimensions, and exacting lifter preload settings, they can perform very well. Harland Sharp, Yella Terra, Competition Cams, Crower, and Crane Cams all make aftermarket aluminum rocker arms for LS-series engines. If you install aftermarket rocker arms and have valvetrain issues, read the next section about pushrods for one of many solutions.

When running the stock rockers is not possible or it just doesn't sit well with you, many companies offer aluminum or steel rocker arm kits. Shown here are the stages of the manufacture of a Harland Sharp aluminum rocker arm from raw extruded aluminum bar, through the machining steps, and into the full-roller rocker arm end product.

Harland Sharp is no stranger to building beefy aluminum rocker arms. It has been building rockers for more than 40 years. Building a rocker arm that works well on the LS-series engines is tricky. They need a really good pushrod and a really stiff spring to operate without valve float symptoms.

One of the steps in building an aluminum rocker arm is pressing the fulcrum bearing into place.

After pressing the fulcrum bearing into place, a snap ring is installed to help hold everything in place. The snap ring isn't required on these rocker arms, but Harland Sharp installs them as an inexpensive fail-safe system.

The LS-series cathedral-port rocker arms are paired by cylinder with a small tie-bar holding them together for assembly. These are typically a bolt-on deal with the correct springs, but installation may require grinding on the valve cover to create clearance because of the rockers' larger-than-stock dimensions.

PUSHRODS

Pushrods don't get much attention when building an engine. Many people just buy upgraded stock replacement pushrods without giving it a second thought, but many don't realize that when you upgrade valve springs and rockers and then run an accelerated-lift camshaft, you may be asking the pushrod to do more than it can handle. If you factor in an open spring pressure of 400 pounds, the pushrod is actually seeing a nonrunning 635 pounds of spring pressure because the rocker arm ratio multiplies the spring pressure back through the pushrod and lifter. The pushrod also has to fight the weight of the rocker arm mass, the actual valve, half of the valve spring weight, and the retainer and keepers while attempting to open that valve. That's a stressful job in the engine. Pushrods have a rough life and get no respect.

Many just grab some 5/16-inch replacement pushrods and throw them in without a thought about what is really going on. You can bet pushrod deflection happens much more than valve float (yet they have similar symptoms), and the longer the pushrod or the higher the rocker ratio, the worse this problem gets. It should be noted that the factory LS7 pushrods, while not hardened, are of a 3/8-inch diameter. GM knows how to make the valvetrain work properly, and if the 3/8-inch pushrod wasn't required on the LS7 engine, it would likely not have been used.

The OEM 7.400-inch long, 5/16-inch diameter pushrods are a weak link when upgrading valve springs, although these can act like a fuse when the engine is over-revved or when it has been mis-shifted, as the pushrod will bend well before the valve in most situations. The LS7 uses factory 3/8-inch-diameter and 7.800-inch-long pushrods.

While weight plays a huge role on the valve side of the rocker arm since it is accelerated much faster than the pushrod, the weight of the pushrod matters much less. You can have a larger diameter pushrod and not affect valvetrain weight much. In fact it's much better to cure and eliminate pushrod deflection than to worry about adding a few grams of weight on the pushrod side of things. Run the largest diameter pushrod you can fit within reason. The only interference normally will come from the cylinder head. You may need to lightly machine or enlarge or elongate the pushrod hole to accept a larger diameter pushrod. Further ahead in this book we will discuss properly measuring pushrod length, preload, and lash, along with further camshaft tech such as degreeing the camshaft and checking piston-to-valve clearance.

When swapping valve springs, camshafts, or cylinder heads, it is highly recommended that you upgrade to a stronger pushrod in the proper length. OEM pushrods have a hard time with upgraded valvetrain components.

For high-rpm usage you may find that you need larger-diameter pushrods or thicker pushrod walls to alleviate pushrod deflection for a smoother operating valvetrain. Use the largest pushrod diameter you can fit, or use a pushrod with more wall thickness if a larger diameter is not a viable option.

TIMING CHAINS

The OEM timing sets are nothing high-tech and are of a simplistic design. It is a basic single-roller timing chain and quite adequate for the job at hand when the engine is left completely stock, but what happens when you start changing engine parts? Higher engine speeds, higher valve spring loads, and aggressive camshaft lobe opening profiles create a much harsher environment for a timing chain. Luckily for us, many aftermarket companies offer upgrades to this sometimes questionable engine part.

The Gen III OEM timing chain is known to be weaker than the Gen IV timing chain. The Gen IV received the stronger C5R timing chain with the introduction of the LS2 engine. While this is still a single chain, the LS2+ chain has much thicker links than its predecessor. It's interesting to note that the LS2+ timing chain can be retrofitted to any LS-series engine. In the case of Gen III LS engines, the chain can be used with the existing timing gear set, making the LS2+ timing chain an economically friendly upgrade. This should be the minimum timing chain upgrade any time you tear into your LS-series engine.

Many aftermarket manufacturers offer single roller chains and many also offer double roller chains. Double rollers are about 50 percent stronger, but knowing what is required and

Throughout the LS-series engine family's life, a few different stock upper timing gears have been available. The top gear is the Gen III timing gear for all engines with rear-cam reluctor sensing. The two front gears are Gen IV specific. At left is a LS2 2x reluctor gear. At right is a center-bolt 4x gear. There is also a similar three-bolt 4x gear.

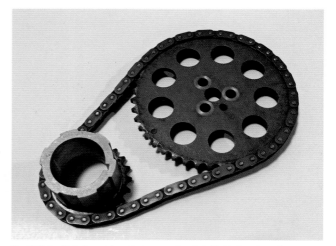

The OEM Gen III timing set is a single-row roller chain that is found on a large spread of vehicle applications. This is the weakest of the LS-series timing sets.

The OEM Gen IV timing set is another single-row roller chain. This comes with the upgraded C5R timing chain, which is commonly known as the LS2 chain.

Comparing the Gen III chain to the Gen IV "LS2" chain, the strength advantages are apparent in the thicker, beefier chain links, which add strength and durability. Notice that the LS2 chain can be replaced while retaining the LS1 Gen III gear set for an inexpensive upgraded chain.

what is overkill will help to make your decisions easier. If you're going all-out with a solid-roller 8,000-rpm setup, the choice is clear: Get the double roller. For the vast majority, a single-roller chain is adequate. There is no harm done in running a double roller chain. You can almost never have too strong of a setup.

One nice feature about all aftermarket chain kits is the ability to adjust camshaft timing. With the factory timing set, you cannot adjust the advance and retard of the camshaft without machining the dowel hole and using offset bushings. With aftermarket sets, there are a few options available to degree in your camshaft to proper specifications. Many timing chain sets use multiple keyways, while a few use an oblong gear that offsets itself over the camshaft indexing dowel. With either setup, it is critical to actually check and degree the camshaft per the camshaft manufacturer's specifications. If you're at or within 1 degree of desired camshaft timing, it's probably best to just leave it alone, in my opinion. Many other factors in play make larger differences.

One thing that has changed between the Gen III and Gen IV LS-series engine is that the camshaft sensor no longer reads off the camshaft reluctor located in the rear of the engine. I may sound like a broken record repeating this reluctor business, but it's important to know which setup you have before finding out you installed the wrong reluctor gears in your new engine. All Gen IV blocks deleted the provision for the rear camshaft sensor in the block itself. When this change was made, GM integrated the camshaft sensor into the timing cover. The camshaft sensor now reads a signal from a matching camshaft gear with built-in reluctor teeth.

The early Gen IV engines used the 2x signal the Gen III LS1 engines used, while the newest designations of Gen IV engines use a 4x timing gear, with four individual trigger points. You absolutely have to use matching crankshaft and camshaft reluctor wheels: 24x crankshafts need a 2x camshaft, and 58x crankshafts need 4x. You also need to use the same reluctor count that the vehicle used when stock. There is no way to change the factory PCM to read a different sensor. You can find out what your vehicle has without tearing into the engine by the color of the crankshaft and camshaft sensors. If they are black, it's 24x/2x, but if either is light gray it's a 58x/4x. The 58x/4x setup will also be on most 2006 and newer vehicles. Keep this camshaft sensor information in your head while you're deciding on which timing chain setup to choose.

CAMSHAFT THOUGHTS

While it's almost impossible to cover all aspects of camshaft designs and theories in a single chapter, I hope you have learned enough to at least not choose the wrong camshaft for your application. No one cam is right for everyone. The camshaft I may like, the next person may dislike greatly. With the LS-series engines still being a "new" engine designation, many cam grinds cover new ground, with no known results other than experience guiding the way. The cam grind (specifically the exact specs) you used to run in your LT1, SBC, or 5.0-liter Mustang will never work optimally in an LS-series engine.

When the single-roller chain is not enough, several companies provide an answer. This is a Rollmaster double-roller timing set with hardened gears, premium double chain, Torrington thrust bearing, and best of all, adjustability for camshaft degreeing options.

Chapter 5
LS-series Engine Machining

The next steps of building your engine involve inspecting, machining, checking, and measuring your chosen components. If you have a new block or a balanced rotating assembly, certain machining steps may by skipped, although verification is still recommended and sometimes needed for peace of mind before assembly time comes. Just because your block is new doesn't mean you can slap it together without checking vital dimensions. Sometimes a used block even needs less work than a new block. Used blocks were likely running engines at some point. New blocks have never been started. Either way, the block will need to be inspected for blemishes, main bore sizing, and the deck height dimensions, in addition to normal machining procedures such as boring, honing, and deburring.

This chapter discusses these topics as well as measuring critical dimensions of the block clearances. If you complete the items in this chapter, your block should be ready for the assembly procedures covered in the following chapter.

Keep in mind this is not a "how to machine" chapter but a discussion of machining topics to inform the non-machinist about behind-the-scenes actions. Knowing beforehand what your block may or may not need can guide and benefit your entire buildup in component choices, block choices, and in who you choose to perform the needed machine work.

ENGINE BLOCKS

Visually inspect the block for cracks, nicks, blemishes, missing bolt holes, or even holes in the block, and anything else that may affect the block's performance. Look over gasket sealing surfaces and where mechanical parts are moving on machined surfaces. If the donor block was a refugee from a catastrophic engine failure, look over these areas even closer and Magnuflux the block to check for small cracks if a cast-iron block is used. If it is an aluminum block, cracking is more evident visually, as the block case will often just crumble away. Same thing goes if the block suffered from water ingestion (hydro-locked). The cylinders with water damage need to have their

Proper machine work technique and experience is what will determine if your engine lasts 500 miles or 50,000 miles. Machine work is best left to the professionals and, ideally, machinists who have had a hand in many LS-series machine work procedures.

Inspecting the block before machining is a valuable step in determining if your block is usable or if you need custom deck-height pistons due to block deck-height issues. Here a block's deck height is checked to determine if it is within the proper range.

liners inspected. It's not uncommon for hydro-locked blocks to have cracks or even broken starter mounting ears. Only use these blocks as a last resort; there is no shortage of donor blocks to build upon.

Another key thing to check early on, before starting the machining, is the block deck height. Deck height (DH) is the distance from the top of the block, where the cylinder head mounts, to the centerline of the crankshaft main journals. DH is important, as it has a direct impact on compression ratios. You want the DH to be close to what your specified compression ratio and somewhat matching what your rotating assembly dimensions are. You also want to make sure the DH on both cylinder banks are within a few thousandths of an inch of each other. Be careful when decking because every amount you cut off the deck directly affects how much more of the piston pops out of the cylinder bores. To see how the DH relates to the rotating

assembly, we need to know a few key measurements that determine the location of the top of the piston. These are easy to procure and calculate. You need to know the half of the crankshaft stroke, actual connecting-rod length (center to center), and the compression height of the piston.

Assuming a 408-cubic-inch LQ4 stroker, we know that half the crankshaft stroke is 2.000 inches (4.000 inches/2 = 2.000 inches) plus the full length of the connecting rod at 6.125 inches and the compression height of the piston (commonly 1.120 inches). The formula to see where the top of the piston would be looks like this: (stroke/2) + RL + CH. Using these numbers we find that this rotating assembly puts the piston height at 9.245 inches. If we use the standard LS-series average deck height of 9.240 inches we can see that the piston is 0.005 inch above the deck surface. If the piston height was 9.235 inches, that is 0.005 inch below the deck surface.

If too much is decked from the block, you may have to either find another block or machine the piston tops down a corresponding amount. It is easier to plan ahead in this case and purchase pistons that have a shorter compression height to compensate. The limit for how far the piston safely comes out of the block is usually around 0.015 inch, as long as the cylinder head itself is not an interference problem. Keep this amount in mind when choosing head gaskets.

The average DH for standard LS-series engines is 9.240 inches. GMPP LSX blocks have extra deck material, which brings the deck to around 9.260 inches before machining. Most builders square this back to standard dimensions by milling the deck 0.020 inch. To find out a rough approximation of where the flat crown of the piston height (PH) will be, you take your piston's compression height (CH on piston spec card), add your connecting-rod length (RL), and add half of your crankshaft stroke. It should be noted that this is the highest point in a true flattop or dished piston, while on a domed or "popup" piston, the dome is not included in CH measurements.

Top of piston crown height = (CH + RL + (stroke x .5))
Example: 408 cubic inches is (1.110 inches + 6.125 inches + (4.000 inches x .5) = 9.235 inches
9.240 inches DH – 9.235 inches PH = 0.005 inch amount below deck piston is at TDC
9.244 inches DH – 9.235 inches PH = 0.009 inch amount below deck piston is at TDC
9.225 inches DH – 9.235 inches PH = 0.010 inch amount above deck piston is at TDC

Actual deck height can either be checked by a machine shop using a CNC machine or by mocking up a partial engine assembly if you have no access to DH measuring equipment. To check DH if your machine shop doesn't or can't do this, mock-assemble your engine with the no. 1, no. 2, no. 7, and no. 8 piston and rod assemblies (no rings are needed, but a degree wheel would be helpful). Measure each piston at absolute TDC compared to the block and write down your readings on all four pistons. There is usually no block taper on OEM or GMPP blocks, so the no. 1 and no. 7 pistons should be close or identical to each other, likewise for no. 2 and no. 8. But if there is a huge difference between the even bank and odd bank, it will need to be corrected. A difference of 0.001 to 0.002 inch is typical and acceptable for a normal street engine, but 0.010 inch is too much and the higher deck height will need to be machined down to match the opposite bank.

The only variable when checking the DH with this method is variance in crankshaft stroke, which can differ by up to 0.004 inch (+0.002 inch to DH) between rod pins at the far end of the spectrum, although generally a less than 0.001-inch variance is the standard.

If you are not using a new block, it's also good practice to check the deck surface for warping using a long machinist's straightedge and a set of feeler gauges (or on a CNC block machine with a dial indicator). The deck surface plays an important part in head gasket sealing. If it's warped or wavy, the low spots will not clamp the head gasket as much as the high spots, and blown head gaskets may result. A variance of

If the deck height is off or the deck is warped, tapered, or corroded, it is best to resurface the deck surface flat. Ideally, the deck height should be set to the same height on each bank when doing this procedure. Using a CNC machine, the deck height can be machined exactly to spec. *Race Engine Development*

0.003 inch across the entire deck surface is tolerable for stock gaskets and mild setups, but in this case less is better. If you can, it is probably best to resurface it and not worry about it afterward. It would be disheartening to assemble an engine only to find out the deck surface needs to be touched up or that the engine needs to be disassembled to do so. Engines that blow head gaskets in the same area over and over again should be looked at closely for a low spot.

Depending on your engine bore size and block type versus piston oversize amount, you may need to have your cylinders bored out. Any quality machine shop will use a torque plate to simulate cylinder head clamping loads while machining the cylinder bores. Typically, if your oversize amount is less than 0.015 inch from current bore size, you can hone the cylinders to size rather than boring them out first. Most aluminum-block-based LS-series setups only need to be honed for your new pistons. The boring machine can be set up to cut out a measured amount from each cylinder. Often a CNC machine is used to bore and deck blocks. The operator can program the machine, walk away for 30 minutes, and come back with one bank of the engine case bored exactly to preprogrammed specifications.

Boring a cylinder can take out a large bulk of material quickly, while honing is a slower and more precise process for removing material and leaving a nice, finished, crosshatched cylinder wall that effectively breaks in the piston rings. The process of boring a block, on the other hand, will leave the block with a rough cylinder wall. Think of it as a long threaded sleeve, not much different than the finish quality and roughness of a freshly cut set of brake rotors. For the old-timers, think of the grooves on a vinyl record.

After boring, these knurled and displaced sleeve and cylinder materials need to be smoothed out. The normal procedure is to leave the cylinder a minimum of 0.0030 inch undersized from the finish bore size, which will leave 0.0015 inch of cylinder honing left to do on each side of the cylinder. The more finish honing required, usually the more precise and cleaner the end product. You don't want the tooling marks left from boring the cylinders to remain, as accelerated ring wear will happen.

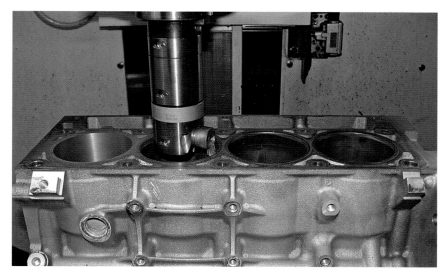

Boring the block is another procedure usually needed with oversized pistons. When boring a block, a good machinist will get it to about 0.004 to 0.005 inch away from ideal bore size and then hone it to size from that point. Boring removes bulk material, while honing removes smaller amounts of material and leaves a crosshatched pattern to assist ring break-in. *Race Engine Development*

These lines are the marks left over from the boring procedure. If left intact or not honed completely, these will mean a quick death for the piston rings. This is why it's best to keep 0.0040 to 0.0050 inches of material left over to hone out, which is really 0.0020 to 0.0025 inch on each cylinder wall. This provides just enough room to completely remove the tooling marks.

When honing a cylinder oversized, the hone operator will start with an aggressive honing stone with 70 to 100 grit for bulk material removal and work toward the final honing grit of 400 to 600, as he slowly creeps up on the desired cylinder bore size. Each finer stone removes less and less material.

The correct honing stone abrasiveness is determined by the recommendations of the honing machine and ring manufacturers. Because the cylinder is honed to a specific size based on piston manufacturer clearance specifications, the machine shop will need your pistons to precisely hone your cylinders to an exact bore size. Aluminum blocks and iron blocks undergo the same basic honing procedure.

A skilled engine hone operator knows what is needed and when. They will operate the engine hone, clean up any cylinder taper, low spots, or high spots, and leave a nice and straight cylinder finish for the new rings to sit upon. Most LS-series engines will receive a final few hone passes using the 400-grit stones. During the entire procedure, the hone operator will verify bore sizing several times.

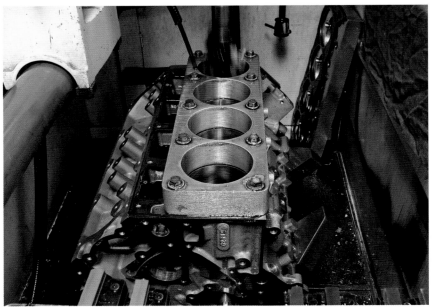

Once boring is completed, the block is honed to the final size. This is done with a progression of abrasive to soft honing stones to produce the final cylinder pattern dictated by ring manufacturer.

During the honing procedure, a skilled machinist will check and recheck bore size repeatedly to creep up on the proper size without going too large. Piston-to-bore clearance is specified by the piston manufacturer and engine application.

After the honing procedure is finished, it's imperative to thoroughly clean the cylinder walls several times to remove abrasive materials. If not removed, this debris will accelerate wear and diminish the rings' useful life. A specific nylon honing brush is recommended to clean the cylinders and smooth the honing scratches from the honing procedure. It cannot be too clean; if you wipe a cloth shop towel down the bore, it should come out clean, with no hints of dark honing stone materials. Before assembly, thoroughly wipe the cylinders with a clean rag using clean solvent or acetone several times.

It's no secret that LS-series engines have extremely sharp machining edges. This is amplified when further machining is done to the block and heads. While inspecting the block, it's a good idea to deburr all sharp edges with either a file, deburring tool, or with a pneumatic die grinder and a careful hand. Oil drain holes can be cleaned up for improved drainage and less restrictive oil return back to the oil pan. You can't always see the benefits of removing casting flash or enlarging oil return holes, but knowing that the block is improved is justification enough. Critical machined edges such as main cap edge surfaces can be lightly filed with a hand-held file or machinist's stone to remove sharp edges. Removing sharp edges and casting flash also has the benefit of removing stress risers where cracks can develop and spread.

ALIGN BORING OR HONING MAINS

The crankshaft main bore alignment and sizing is often overlooked when building an engine. The bearing bore sizing should not vary by more than 0.0001 inch between any of the main journal bores. If there is a discrepancy in sizing, this can be corrected by reboring or honing the block. Another reason to align bore or hone would be the addition of main cap studs or do it when installing forged and billet main bearing caps. The added torque from studs will distort the bearing housing.

If installing main cap dowels to keep main cap movement in check, the block will also need to be align honed. Basically anything veering from the stock setup needs to be checked and dealt with properly.

When align boring or honing, the stock main caps need to have the flat clamping surface milled slightly so that a completely new bearing bore diameter may be cut. The only drawback is that the more you need to correct an issue, the closer the crankshaft centerline moves toward the camshaft centerline. Too much of a move can cause timing chain issues.

CAMSHAFT BEARING REMOVAL AND REPLACEMENT

The camshaft on an LS-series engine resides in the block above the crankshaft and rides on five full-circle engine bearings that are designed as one piece. These bearings are a press fit into the block and usually need to be knocked out before machining a block, unless the block is only getting honed. During honing, the block rests on the main bearing bore. The bearings have a 0.008- to 0.009-inch interference fit into the block. A special camshaft bearing tool is required to remove and reinstall these bearings. The cam bearings have a unique external size at each location, 1 to 5, and need to be in correct orientation, in both location and proper alignment with oiling holes to lubricate the camshaft main journals, before being pressed into place.

In aluminum LS-series blocks, a sleeve retention compound can be used to "lock" the new camshaft bearings in place. Aluminum blocks expand greatly with engine heat compared to iron blocks, so while sleeve locking compound isn't required with either, it is a good idea to use it. The philosophy is similar to the idea of using thread locking compound on critical bolts.

Once the machine work is finished, the machine shop can install the camshaft bearings. These are easy to do with the right tools. The first step is to line up the oiling hole with the slot in the block. If you do not have a cam-bearing installer, don't bother trying to do this yourself. Let your machine shop install these.

On an OEM block, you can simplify the installation alignment by drawing a line directly between both oiling holes in the camshaft bearing. One is used and one is not. Using the line method puts one of these oiling holes into the proper location. Note that there are three cam bearing outside diameters. Mark them and install them in the right locations.

With the cam bearing in its proper orientation, the bearings can be knocked into position using the cam bearing installer. The only hard part here is making sure the depth is correct. Use a 1/8-inch, 90-degree Allen wrench to verify the oiling hole is 100 percent lined up to the block oiling passage.

After installing new camshaft bearings, a test camshaft should be installed to check cam bearing alignment. Any LS-series donor camshaft will work. Tape up the camshaft lobes with masking or electrical tape to protect the newly installed camshaft bearings from damage. You are just checking to make sure the camshaft physically fits and that it is not binding during rotation. The reason to do this before short-block assembly is that it is much easier to repair a camshaft bearing issue now, rather than when the rotating assembly is in the block and directly in the way.

The camshaft bearing clearance is often overlooked. Many people, including machine shops, slap these together without giving cam bearing clearance a second thought. If it's too large, the cam can become an internal oil gusher, leaking all your oil pressure internally. You measure the OD of the cam journal first.

After taking your cam journal OD measurement, set the dial bore gauge to this size and measure the corresponding oil journal clearance. Oil clearance typically will be in the 0.002- to 0.004-inch range, although with aftermarket bearings this can be as loose as 0.006 inch. The more toward the loose side of the scale, the lower the oil pressure may be due to stacking clearances (if more than one is looser). If this is too large or small, it is necessary to correct this issue before going further.

If oil journal clearance is ideal, the next step is verifying camshaft bearing alignment. It is not hard to have one bearing not aligned to the rest. Sometimes they get misaligned while you're installing the camshaft bearings. Use a test camshaft with the lobes taped up to test and check installation.

The camshaft journal oil clearance can be checked at this point, although it rarely will need attention. Many engine builders will just verify that the camshaft doesn't bind or isn't too sloppy and that's about all that is needed. Camshaft bearing clearances are similar to crankshaft journal clearances and are specific to the block material type. If the bearing clearance is physically too tight, there are a few options. The easiest may be as simple as locating a different brand of camshaft bearing, but if this is not an option, the camshaft journals may be reground and polished to size. If the bearings are too loose, a brand change may be enough. In the worst case you may need a different camshaft with slightly larger journals, although this is a rarity.

Similar to checking camshaft bearing clearance, we also need to check the highly critical main bearing clearance. This clearance varies according to block material and application. The first step is to measure each journal in at least two spots and note the sizing.

While the micrometer is out, also check the connecting-rod journals. The target clearance does not vary between aluminum and iron blocks. Measure in at least two spots on each journal for each bearing location to determine accurate sizing.

MEASURING AND ADJUSTING MAIN BEARING CLEARANCE

Aluminum and iron LS-series blocks have different needs and conditions for bearing clearances. Aluminum has almost twice the expansion rate per inch of material than an iron block. Iron-block LS-series engines are similar to any other iron-block engine for bearing clearances, usually close to 0.0010 inch per inch of bearing journal diameter. Iron block main clearance is always looser than aluminum blocks when at room temperature. Iron blocks will often be set 0.0005 to 0.0010 inch looser than the same setup in an aluminum block. A typical iron block main clearance will be 0.0025 to 0.0030 inch, while aluminum-block engines need to have 0.0015 to 0.0020 inch of clearance. Power-adder setups should be considered when setting main bearing clearance, as its better to be slightly too loose than tight with any setup. The larger the clearance, the lower the oil pressure will be, as bearing clearance is the restriction that gives engine oil its pressure. Because high-pressure oil always travels to low-pressure areas, any significantly loose bearing can bleed off oil pressure.

There are a few ways to check main bearing clearance. One of the key things to remember is to always check clearances with all components at room temperature. That includes measurement tools as well as the parts you are measuring. Don't leave your block or crankshaft outside in the sun or in a freezer before making these measurements. These are both extreme situations, but it's important to have all engine parts sit in a controlled environment at least overnight before measuring. If you have access to a dial bore gauge and a set of micrometers, this is the best and most accurate way to check clearances.

The procedure is to measure and document the dimensions of the no. 1 to no. 5 main bearing journals of the crankshaft you are using in your buildup with a 2- to 3-inch micrometer and then zero out your dial bore gauge to that exact measurement. Then install the main bearing inserts in the block in the proper locations. The insert with the long oil groove side is the upper block side, while the smoother bearing is the lower main cap bearing. Install the main caps and torque to specifications. With ARP main studs, torque the inner nuts to 60 lb-ft and the outer nuts to 50 lb-ft (stock inner bolts to 16 lb-ft, then an additional 80 degrees rotation; stock outer bolts to 16 lb-ft, then 51 to 53 degrees additional rotation using an angle torque measuring apparatus).

With the bore gauge zeroed out, measure the corresponding main bearing clearance against the same crank journal with which you zeroed out your gauge. Positive numbers will indicate bearing oil clearance. You will have to move the dial bore gauge around a little to find the loosest spot, and that is your oil clearance reading. If your measurement is off the scale to either side, too tight or way too loose, re-mic the crankshaft journals, reset the dial-bore gauge, and then recheck the measurements again until you are confident in your measurement accuracy.

Check three spots on both the front and rear sides of each bearing to verify and document these numbers as you measure them. Your local machine shop most likely can take these same bearing clearance measurements if you supply them with the rotating assembly and pay the labor for their measurement time. Document all measurements on an engine build sheet for future reference.

Continued on page 92

TORQUE ANGLE MEASUREMENTS AND TORQUE-TO-YIELD BOLTS

Angle torqueing fasteners may be something new to many, and often it may be hard to understand why angle torqueing is necessary, when torqueing, to a specific torque setting is a much easier procedure. With angle torqueing you are relying on specific and accurate bolt stretch clamping. All angle torqueing techniques will have a low base torque setting (16 to 22 lb-ft), which promotes tightening up the bolt to a specific unstretched "starting-point" location. The second step after the base torque setting will usually be an angle degree reading (0–360) with each 90 degrees being represented in one-quarter-turn increments. Since the manufacturer knows the thread pitch of the bolt, the bolt stretch amount is easily calculated by the amount of turning the bolt is moved during this step. Bolt stretch is more accurate at clamping loads than using a torque wrench to torque the same bolt. This is due to unmeasurable friction influences when turning the bolt to the torque specification.

Angle torqueing and torque-to-yield are two different things, depending on the bolt torque. An example would be the LS-series OEM main cap bolts. While these are angle torqued during the tightening procedure, they are not torque-to-yield and are reusable a few times; whereas, the OEM LS-series torque-to-yield head bolts are permanently lengthened each time they are installed and torqued when new. The torque-to-yield bolts must be replaced with each use, while angle-torqued bolts do not require it most of the time, unless damaged or overtorqued.

Dry assemble the block with the main bearings in place and torque the main cap bolts to specifications.

OEM fasteners use a torque specification and then an angle measurement that uses bolt stretch rather than friction to determine proper tightening. All bolts are torqued to 16 lb-ft. Then the inner bolts are tightened in paired to 80 degrees. Then the outer bolts are turned 53 degrees.

Set the dial indicator to correspond to your journal outer diameter measurements and measure each corresponding main journal bearing location. Iron and aluminum have different needs for actual main bearing clearances.

Just like the main bearings, we want to check the rod journal oil clearance. Make sure to torque the connecting rod caps to recommended specs before measuring, as incorrect torque values will distort the cap differently.

If you don't have access to a dial bore gauge, there is another way to check clearance, although professional builders would use this more as verification if the dial bore gauge measurement is questioned. This method isn't as accurate, but it is close enough to show if there is a problem that needs to be corrected before moving on. Clevite offers a Plastigage strip, which is a very thin piece of waxy thread that you use after installing the crankshaft in the block. For LS-series engines, you need the "green" 0.0010- to 0.0030-inch range Plastigage to measure the crank main and rod journals. Put a thin strip on each clean crankshaft journal, and torque down the crankshaft main caps. Then carefully remove each main cap and measure each compressed Plastigage thickness after being squished between the bearing and crankshaft journal. Compare this width to the paper included with the Plastigage, and it will tell you the bearing clearance. Accuracy seems to be within 0.0003 inch, which is plenty accurate for most homebuilt engines. This same method will work for connecting-rod bearings as well.

The engine should be upside-down with the crankshaft weight resting on the upper main bearings when checking main bearing clearance with a Plastigage. During measurements, do not use assembly lube if possible. Dry or lightly oiled bearings will provide the most accuracy. A thin film of oil is all that is needed for now. Save the thick assembly lube for final assembly. To be more accurate with rod bearings, push the piston and rod assembly against the crankshaft by hand to make sure the rod cap bearing sees the slack end of the rod bearing, where clearance is greatest. If clearance numbers are suspect from this method, recheck again to verify. You can also rotate the crankshaft 180 degrees to check both sides of the crankshaft journals for further verification. Make sure to thoroughly clean the Plastigage from the crankshaft and bearing surfaces. A small plastic scraper will remove the Plastigage without harming the engine components.

What do you do if you have a bearing clearance issue? Many bearing manufacturers produce bearings that are 0.001 inch tighter or looser than the standard sizing. This can be a mix-and-match game, as sometimes one journal will be loose while all the others will be normal. You can replace the individual bearing that is loose. You don't need to use the whole set of +/- bearing sizes. If you buy two bearing sets to do this, you will have one mismatched bearing set left over that is probably useless unless you are building more than one or more engines. It should be noted that manufacturers have some slight variances in dimensions, so getting the right clearance sometimes can

If access to a dial bore gauge is impossible, then another way to check oil clearance is the use of inexpensive Plastigage. It is sufficiently accurate for most engines. You start by laying a strip of Plastigage on the dry journal surface after assembly.

Next, carefully install the bearing rod cap and torque to specifications. This will crush the waxy Plastigage thread in the small gap that is left, effectively showing the bearing oil clearance.

Using the paper sleeve that the Plastigage comes in as a guide, line up the squished wax to the sizing chart. The line where it best matches indicates the bearing oil clearance. This one was more than 0.002 inch and less than 0.003 inch, so you can see that this isn't an exact measurement. Use a piece of soft plastic to clean off the wax.

be a matter of changing bearing brands. If the block is align honed, bearing selection is a little easier and more uniform throughout the bottom end of the engine, depending on crankshaft journal sizing and manufacturer quality control at that point.

STROKER CLEARANCING

With the additional stroke from 4.000-inch, 4.100-inch, and 4.125-inch crankshafts, sometimes block interference rears its ugly face during your engine build. It's always a good idea to mock assemble your block before installing rings and check the block clearance to make sure there are no interference areas that need to be ground out. With connecting rods and pistons moving at a high rate of speed, you can't have "just enough" clearance where the rod bolt barely rubs the block. If you can't fit a sheet of paper between the close areas, you need to spend some time fixing that. On the rotating assembly, you want to have at least 0.050 inch of space anywhere components get close to touching. Usually this ends up being a rod bolt coming too close to the bottom of a sleeve, depending on connecting-rod brand. This can be remedied by marking the offending area with a black marker and then disassembling the block and grinding the problem areas until proper clearance is achieved.

Some clearance issue areas are hard to physically check with a feeler gauge, so a visual inspection may be all the indication of proper clearance you can achieve. Think of the spark plug gap when checking these tight clearance areas. If you've got a space at least the thickness of a stock spark plug gap, you're good to go.

Many machine shops and engine builders already know where the close areas are and will pre-grind them before assembly to save some time. This is known as stroker clearancing. Sometimes this is done on a CNC machine rather than by hand, but it is quicker and cheaper to do it with a hand grinder for the same end result. If cutting into a sleeve, care needs to be given that just the right amount is removed. After grinding, a deburring process is required if the sleeves have already been machined for the new pistons. If the stroker clearancing is done before honing, the honing process itself will deburr these areas. It's good to have a small radius at the bottom of the cylinder wall so that the machined edge of the sleeve does not catch on the piston like a razor's edge as it shifts from downward movement to upward movement at bottom dead center.

If not previously done with the machining you may find that some of the connecting rod bolts may hit the bottom of the cylinders when you start assembling the engine. There is an easy though messy fix. Mark these areas with a marker and then just grind enough to clear. Once clearance is adequate, make sure to deburr and chamfer the bottoms of the cylinders if sharp edges exist.

COOLANT AND OILING SYSTEM PLUGS

New engine blocks do not come equipped with external plugs, dowels, or oiling system plugs. These obviously need to be installed at some point, and the best time to do this is before short-block assembly. The threaded engine plugs may be reused from your old block, but they are also inexpensive from GM, so it's your choice. There are two coolant plugs: one large coolant plug is on the driver's side, which takes a 17-millimeter Allen wrench, and the other is above the starter location, which takes an 8-millimeter Allen wrench. Before going much further, make sure the block's oil gallery holes are free from debris, whether machining chips, bugs, or a blob of dried up silicone. It's a good idea to run a pipe cleaner through the oiling holes and then running clean compressed air through these lines before assembly.

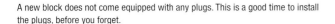

A new block does not come equipped with any plugs. This is a good time to install the plugs, before you forget.

There are two different designs of rear barbell restrictors. The front design is the original design, which was prone to leaking oil pressure under the rear cover because of the O-ring placement. The newer design alleviates this issue simply by relocating the O-ring further in.

It is easiest to install the barbell restrictor before putting the engine on the stand. I was lucky enough to have just enough room on this engine. Use some assembly lube on the O-ring so as not to cut it.

With the O-ring portion facing toward the back of the block, install the barbell restrictor by pushing it into place. They sometimes can be pushed into place by hand. If you find it difficult to do this, a small hammer can be used to assist.

Regarding the oiling system, there are two external oiling gallery threaded plugs on the driver's side of the engine, and both are 8-millimeter Allen plugs. There are additionally two internal plugs that greatly affect oil pressure if left out. The internal plugs and two most critical is the small core plug in the front and the plastic barbell restrictor in the rear. Without either of these internal plugs, you will absolutely have zero oil pressure. The front or rear engine cover would need to be removed to install these after the engine is installed, after you get everything else out of the way.

These are easy to install. The core plug can be installed with the back side of a ball peen hammer, and the plastic

barbell restrictor can be installed by hand with a light oil coating on the O-ring. Do not reuse either of these internal plugs from your old engine. GM redesigned the plastic plug depth to not pop out under engine oil pressure, as some early engines had internal oil leaks past the barbell O-ring portion. It's amazing how many people forget to install these two plugs in their new GM blocks. Don't be that guy. Put in these plugs well before short-block assembly starts, and double check for them on any engine before installing the timing cover and rear engine cover. Always assume all blocks do not come with these plugs, and have them on-hand just in case.

The front oil core plug can be installed using a punch and hammer, but the easiest way I have found has been to place the back-end of a ball peen hammer against the core plug and use another hammer to strike the first hammer, knocking the core plug into the perfect depth and placement.

Here the front core plug is installed completely. This can also be replaced with a shortened 3/8ths threaded pipe plug for a permanent installation. As with any drilling, tapping, or grinding, make sure to wash out the block and use compressed air to keep out debris.

Next, install the large threaded Welch plug into the left cooling system drain hole. This takes a rather large 17-millimeter Allen wrench to install.

Proceed to the three external threaded engine plugs for the cooling and oiling system. There are two plugs on the driver's, or left side, and a single coolant plug on the passenger's, or right-hand side. Torque these to 44 lb-ft. Use liquid threat sealant if you are reusing donor plugs.

When building the GMPP LSX block there are a few extra plugs that the OEM blocks do not have, such as extra cooling system Welch plugs. You'll need an extra block plug to cap off the priority-main drilled oiling hole under the rear engine cover. This uses a 3/8-inch pipe plug using a 5/16-inch square drive bit. This plug seldom can be installed fully. If it's not seated fully, the rear engine cover will not seat correctly and it will gush oil from the rear of the engine. When installing this plug, make sure to test fit the rear cover for adequate clearance. Often you'll need to grind the plug shorter, or you'll have to purchase a shorter Allen-headed pipe plug as an alternative.

With the bare block in completed form, bearing clearances checked, coolant plugs installed, and all block-related machine work finished, coat the cylinder walls with a film of clean engine oil to prevent rusting in humid environments and to keep dust out of the engine while it awaits short-block assembly. It's a good idea to wrap or store the engine in a sealed engine bag. Even a large trash bag will do the job if nothing else is available. Just make sure there are no leaves or yard trimmings in the bag.

When building the GMPP LSX block, this threaded oil gallery plug needs to be blocked off before installing the rear engine cover. It takes a 3/8 NPT plug, which often needs to be machined shorter than supplied due to interference with the LSX rear cover. Notice the camshaft journal oiling hole is at 6 o'clock on these LSX blocks; whereas, it is angled on the OEM blocks.

Chapter 6
Short-Block Assembly and Setup

Finally we're at the point where some real work can start happening. In this section we'll be covering the short-block assembly procedures with some tips and tricks along the way. Building a performance LSX-based engine assembly is nothing like rebuilding the 350 SBC in your old man's 1982 Chevy pickup. I know we've all seen someone slap together an engine in an hour or two, without measuring or verifying anything in the buildup.

These are the most critical aspects of engine building, however. In the previous chapter, we measured and documented crankshaft and camshaft oil clearances. We touched on cylinder wall preparation and measuring bore. This chapter will deal with measuring and assembling the remaining short-block components. When finished, we will have the short-block finished. We're doing this in stages as there are a few variances in different builds that need to be covered in a detailed fashion.

To build any performance engine, necessary procedures must be followed to get from a pile-o-parts into a precision-built and -assembled short-block as pictured here.

The first order of business is to look over the block to make sure there are no problems and that the machine work you paid for was done correctly. Wipe down the cylinder bores and clean out oiling journals throughout the block with acetone solvent or spray brake cleaner and dry compressed air.

GETTING STARTED

Because there are so many parts in a short-block, you can almost start anywhere, but we'll go in order of assembly. We'll be starting with installing the balanced crankshaft, then assembling the piston and rod combo, file-fitting, and measuring piston ring gaps, engine assembly, and finally measuring assembled clearances. If your block is not already installed on an engine stand, now would be a good time to get it ready. Remember that even the lightest aluminum block is 110 pounds, so use an engine hoist or an overhead crane to safely lift the block high enough to bolt the engine stand to it.

CRANKSHAFT

We previously discussed measuring clearance on the main bearing journals, but before the crankshaft is installed, the rod journal sizes also need to be measured with a micrometer and documented on the engine build sheet if not previously checked. These measurements are used to verify and inspect oil clearance on the connecting-rod bearings. Once all measurements are taken, the crankshaft can be cleaned in a parts washer, or with some spray solvent in a can. Clean out all crankshaft oiling holes thoroughly, and clean all rust-protecting oil and grease from the crankshaft journals. The crankshaft should be somewhat clean already. If it was balanced at your local machine shop, it would have already been cleaned once if any drilling or lightening was performed during balancing. You can never be too careful. Run clean solvent through the oiling holes and blow them and the crankshaft surfaces clean with compressed air before installation.

Next, remove the block's main caps in preparation for installing the crankshaft. The properly measured and sized main bearings should still be in place from the previous chapter's machining and measuring procedures. If you have ARP main studs still installed (which should have been installed before align-honing the block), you need to be extra careful when guiding the crankshaft over the studs so as not to knick or damage the bearing surface. Get help from a buddy to set the crankshaft into place to avoid damage. The crankshaft will normally weigh about 50 to 52 pounds. Removing the studs or placing rubber vacuum caps and hoses over the studs would be some cheap insurance if you are unsure of your crankshaft wielding skills.

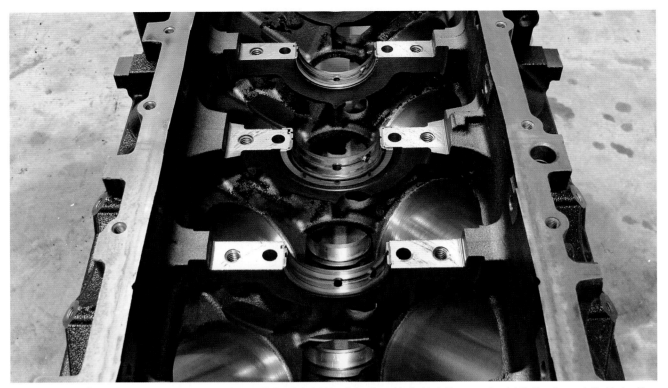

With the engine block cleaned, we can now start short-block assembly. Install the initial main bearing set in the block. The grooved and slotted bearings go in the block, with the center bearing being the thrust bearing. Remember to align the bearing tangs into the machined slot.

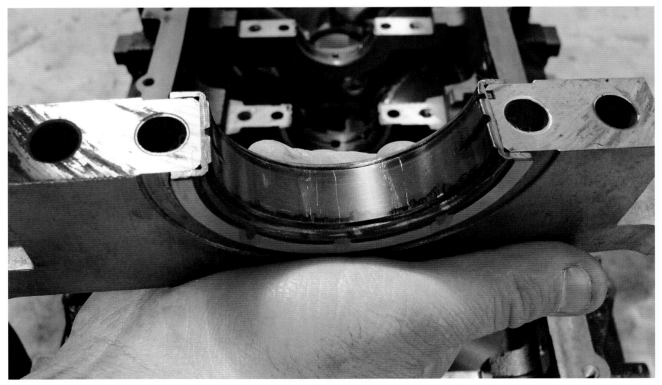

Now install the matching smooth bearing in the corresponding main cap locations. The no. 3 main cap gets the opposite side thrust bearing. Some main caps have bearing tangs machined in either direction. We want to make sure these are installed on the opposite side as the upper block-side bearing.

Before setting the crankshaft into place, use your favorite engine assembly lube on the bearing surface. I recommend and have used Redline Engine Assembly Lube and Permatex Ultra-Slick Assembly Lube with good results. Don't be worried about using too much. It will push out if you get too crazy with it anyway. Lube both the upper and lower main bearings liberally. Now gently and carefully set the crankshaft into the block. I don't like to spin the crankshaft over until the main caps are tightened down a little and after clamping the bearings into place. On OEM blocks the no. 1 to no. 4 bearing caps are stamped "1" through "4" on the driver's (left) side of the main cap, while the no. 5 main cap is flipped 180 degrees and marked on the passenger (right) side of the engine. It is reversed to clear the rear engine cover. All main bearing tangs are also 180 degrees apart from each other.

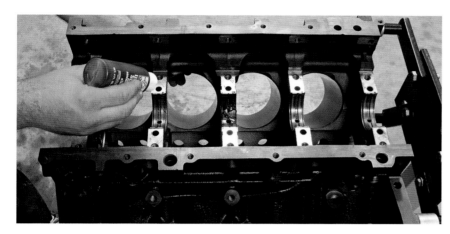

Using your favorite engine assembly lube, evenly distribute it over the bearing surfaces. Don't forget the forward and aft surfaces of the thrust bearing. The assembly lube provides adequate lubrication upon startup until engine oil pressure takes over.

If using Automotive Racing Products (ARP) main studs, it is best to remove them for the crankshaft. If you decide not to, you can proceed by laying the crankshaft into place, paying heed to the crankshaft reluctor wheel. Gently set the crankshaft into place and do not move it until the main caps are torqued.

Now reinstall the ARP studs using clean oil on the threads that protrude into the engine block. Do not tighten these studs more than hand-tight. The studs are a similar total length, but the outer studs have a longer threaded portion for windage tray mounting.

Place the main caps back into place, using a wooden hammer handle or rubber mallet to softly tap the main caps into place. It's not good practice to pull the main caps into place with the threaded bolts. Starting in the middle, snug up the main stud bolts or nuts. If using stock hardware, you need to put a few drops of oil on the threads and under the head of the bolt for proper torqueing. If using ARP hardware, use ARP's recommended moly lube on the threads of the stud and on the washer face of the ARP nut. Before torqueing the main bolts, you need to align the two halves of the crankshaft

Just like the upper crankshaft main bearings, apply your favorite assembly lube to the lower bearings in preparation for assembly.

Alternatively, you can lubricate the main bearing journals instead of or in addition to the main bearings themselves. Gently guide each main cap into proper position and orientation. The numbers stamped on the no. 1 to no. 4 main caps are aligned to the odd-numbered cylinder side, while the no. 5 is flipped opposite. Forged LSX and LS7 main caps have alignment arrows indicating orientation towards the front of the engine.

thrust bearing, which is located in main cap no. 3 right in the center of the engine. Tap the crankshaft forward with a rubber mallet or soft-face dead blow hammer, then tap the crankshaft backward once and forward once more before torqueing the bolts. You can also apply light forward pressure against the crankshaft with a small pry bar while torqueing the no. 3 main cap bolts. Aligning the two thrust bearings allows both half-bearing thrust surfaces to support the crankshaft equally in either direction. You may notice that the crankshaft will rotate more freely in the block when aligned.

After positioning the main caps into proper orientation, they can now be gently tapped into place. You should not need much force to tap them into position.

If using the ARP main stud kit, you will have 20 matching ARP nuts. Early kits also include hardened washers. A mild coating of ARP moly lube is required on flat friction surfaces and threads to obtain accurate torque values.

The torque sequence starts in the center with all the inner bolts first, working outward in a circular pattern until all bolts are torqued down. If using OEM main cap fasteners, keep in mind that you are slightly stretching these bolts during torqueing. While it's not a requirement to replace them for each use, many builders (myself included) only like using them once per running application of the block. They do not get stretched as much as the throwaway OEM head bolts and are a better quality bolt. Time and retorqueing seem to take their toll on the factory main bolts. If you have a used donor core block and used bolts, my recommendation is to replace the fatigued bolts with new GM fasteners or purchase the reusable ARP main stud kit. If you are using a new block, it will come with new fasteners.

After installing all ARP nuts or OEM fasteners, torque to 15 lb-ft and align the thrust bearing by first firmly tapping the crankshaft forward with a soft-faced hammer.

Then firmly tap the crankshaft rearward and then once more forward. This thrust bearing alignment process aligns the two flats of the thrust surfaces so that they equally support the crankshaft thrust surface and provide proper thrust clearances.

Now that the thrust bearings are aligned we can proceed to the regular torqueing process. While maintaining forward thrust load, torque the inner main ARP studs to 60 lb-ft. Once the no. 3 thrust main cap is torqued, you do not need to force the crankshaft forward.

If using an ARP main stud kit, apply a mild dab of ARP moly lube to the stud threads and on the washer face of the ARP nut. If using an early ARP stud kit that uses a hardened washer also, coat both sides of the ARP washer with moly lubricant as well. Using a 12-point 12-millimeter socket on the early kit, or a 12-point 11/16-inch socket on the new kit, the ARP inner main nuts are torqued to 60 lb-ft in sequence and steps, increasing the torque setting each time around. The outers are torqued to 50 lb-ft in sequence and steps too. Next, the 10 M8 external side main cap bolts are torqued to 19 lb-ft. The ARP external side main cap bolts need RTV sealant applied to the washer surface before installing because a lack of sealant will cause oil to seep to the outside of the block past these bolts.

ARP Main Stud Torque Specifications (follow rotational pattern)

1st step inner = 30 lb-ft
2nd step outer = 25 lb-ft
3rd step inner = 60 lb-ft
4th step outer = 50 lb-ft
5th step side M8 bolts = 19 lb-ft starting in center working outward

Tighten all inner main ARP studs to 60 lb-ft following a center-to-outward pattern torqueing both bolts on each main cap in pairs. Then tighten the outer main studs to 50 lb-ft following the same pattern working from the center outward.

If you are using OEM fasteners, these use a base torque, then an angle torque procedure. First, torque all bolts to 40 lb-ft and loosen to no less than 4 lb-ft to initially seat the main caps. The inner main bolts are torqued to a 16-lb-ft base value and then 80 degrees farther, working from the center outward in a circular pattern. Remember to maintain forward thrust on the crankshaft.

Torque the outer OEM studded main bolts to an initial setting of 16 lb-ft, and then add 53 degrees of additional tightening.

The factory bolt torque specifications use a base torque setting and then angle measurements to deliver consistent bolt stretch amounts. Before going all willy-nilly and jumping in to torque the bolts, remember to set the thrust bearing alignment by either tapping the crankshaft forward, rearward, and back forward with a soft-faced hammer or by maintaining forward load on the crankshaft while torqueing the bolts. All the bolts must initially be pre-torqued to 40 lb-ft to seat the main cap into place, then all may be loosened to no less than 4 lb-ft (48 lb-in). The inner stock bolts are first torqued to 16 lb-ft and then turned an additional 80 degrees. The outer stock studded bolts are also torqued to 16 lb-ft, but they are only angle torqued an additional 51 to 53 degrees, as they are shorter than the inner bolts. All told, the inner bolts are stretched approximately 0.017 inch, while the outer bolts are stretched at approximately 0.011 inch.

New GM side bolts are equipped with an orange sealer applied and need no additional sealant if new. These bolts are reusable, but in this case RTV sealer is needed to prevent oil seepage. To be safe either way, a dab of sealant won't hurt. Torque the side bolts to 18 lb-ft.

Factory Main Bolt Torque Specifications (follow rotational pattern):

1st step = 16 lb-ft
2nd step inner M10 bolts = 80 degrees +/- 3 degrees
2nd step outer M10 bolts = 51 degrees +/- 3 degrees
3rd step side M8 bolts = 18 lb-ft starting in center
    working outward by pairs

Whether reusing GM side main cap bolts or using ARP bolts, a dab of black RTV silicone sealant should be used under the bolt face to prevent oil leaks. New GM side bolts have an epoxy-based sealant that works well and requires no additional sealant.

Install all side main cap bolts by hand first, and then torque by pairs to 18 lb-ft. ARP specs call for 19 lb-ft. If using RTV sealant, make sure to torque the bolts immediately after hand-tightening them.

The crankshaft now should easily spin by hand pressure only. You shouldn't need a wrench and lever system to turn the crankshaft by itself. If it doesn't turn easily by hand, something is wrong. If you have a locked-down crankshaft at this point, I recommend that you loosen one main cap at a time, and when the crankshaft is able to freely turn again, look over the last main cap you loosened closely for any mistakes. If proper machine work was done, and bearing clearances are spot-on, there should be no issues. Usually if a problem like this arises, the main caps are on backward, and it's a simple fix to alleviate the issue.

If the crankshaft rotates and no problems or interference issues arise, you may now check crankshaft thrust clearance. This is the measurement of crankshaft movement front to back along the length of the crankshaft as it sits installed. Using a dial indicator on a fixed surface (the engine stand where it bolts to the engine itself is adequate), you can pry the crankshaft forward and rearward, checking dial indicator high and low numbers, measuring on the rear crankshaft flywheel flange. With this measurement, you can tell what the crankshaft thrust clearance is and if it needs attention, or if it's alright to move on. The OEM service specifications call for a range of 0.0015-inch to 0.0078-inch on crankshaft endplay, if it is too loose not much other than replacement or repair can be done with the crankshaft.

If the clearance is too tight, regrind the thrust surface on the crankshaft a touch more, or slightly remove bearing thrust material with an extremely flat surface and 220-grit or finer sandpaper to loosen the clearance a minute amount. Both scenarios are uncommon when using new components or reusing parts that had no prior issues.

PISTON AND CONNECTING-ROD ASSEMBLIES

Set the engine aside for now and cover it with an engine bag to seal out contaminants. We have some time to spend setting up the piston and connecting-rod assemblies if they are not already assembled. The piston and connecting-rod component descriptions are found in Chapter 2, Rotating Assembly. Here we'll cover installing the rods onto the pistons properly, installing piston pin locks, setting ring gaps, and finally installing the piston assemblies into the engine block. There is no set order. You can file-fit rings first or last, as long as it's before final assembly of the short-block.

The factory rod and piston combination had all eight rods and pistons, which have no valve reliefs, assembled in exactly the same direction. Of the engines that do have valve reliefs (7.0-liter LS7, 6.2-liter L92), the pistons are still installed in the same direction. The portion of the piston that faces the front of the engine is marked with a dimple. With stock engines, all eight pistons are the

Before getting too far ahead of ourselves, we need to check the main thrust bearing clearance. To do this, use a dial indicator on a secure point of reference. This can be the engine block or the engine stand. Once the dial indicator needle is against the crankshaft, pry it forward and back, noting the amount of change.

same identical part, while on aftermarket pistons, they are grouped by engine bank. Left bank cylinders 1, 3, 5, and 7 are one piston design, and right bank cylinders 2, 4, 6, and 8 have another piston design. Piston no. 8 is unique to itself, as some manufacturers remove material from the piston pin boss to clear the crankshaft reluctor wheel. This is usually dependent on piston pin length. If you can see the offset visually, the piston pin is offset to the major thrust side of the block, which is always the passenger side (right side) of the engine block with the block upright as installed in the vehicle. If your pistons are marked "F" for front, this becomes a little easier, as you may match up the pistons in groups. If all eight pistons are laid out on a table with the piston tops facing up and valve reliefs lined up, four will have the "F" location on one side, and the other four will have the "F" on the opposite side of the piston.

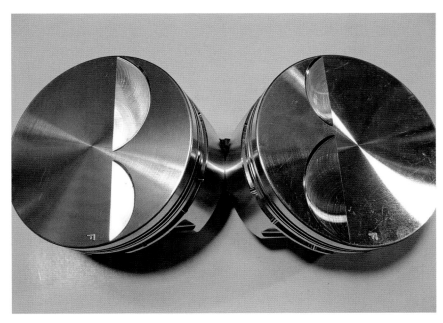

If the pistons and rods are not married yet, we need to help them out a bit. When using pistons with offset piston pins, there are two differently machined pistons. Left and right banks each have their own piston. The "F" always goes to the front, and the valve reliefs reside on the intake side of the block. They only go on in one correct position. Alignment is crucial to engine longevity.

In addition to the offset piston pins differentiating each set of four pistons by engine bank, the no. 8 piston will sometimes need a machined reluctor pass for those pistons that use 2.5-inch long piston pins. This piston needs to be mated to the no. 8 cylinder and rod location.

Connecting rods are all identical in the set. The easiest way to quickly remember which direction to install rods is normally by noting the rod bearing tang location. With stock rods, which are all installed in the same direction, all rod bearing tangs will face the driver's side of the engine. On all aftermarket rods, the bearing tang will face away from the centerline of the engine. Even numbers face toward the passenger side of engine, while odd numbers face the driver's side. Aftermarket connecting rods and matching rod bearings have a chamfer that clears the radii found on the connecting-rod journals of all aftermarket forged crankshafts. When installed in the engine, each rod pin will have two connecting rods, and the chamfer will be 180 degrees apart between the two connecting rods. The forward connecting rod on each rod journal will have the chamfer facing the front, while the rearward connecting-rod chamfer on that same journal will face toward the rear of the engine. The chamfer always faces the crankshaft; if the chamfer is against another connecting rod, it's wrong.

The connecting rods should be numbered from the balancing procedure, but if they all weigh exactly the same, numbering matters a little less. It's always good practice to keep parts numbered throughout engine-building procedures, especially if you're tearing down an engine intended to be reassembled or rebuilt. Previously run parts tend to have certain wear characteristics, and these parts need to be put back into their same locations.

Before assembling the pistons and rods, look over all parts to make sure the piston pins physically fit and match the pistons and connecting-rod size, both in diameter and length. If all parts match and are correct, you may move on to piston and rod assembly. Install one piston pin retainer (spiro lock) on one side of the piston; which side does not matter. You do not need the rod or pin right now. There is a small learning curve to installing spiro locks. The first few are tricky to install, but remember if they are hard to get installed, you should feel much better about them staying in place at 8,000-plus rpm once installed. The good news is there is a very low failure rate due to spiro lock's falling out. Some pistons are designed to use a single spiro lock on either side of the piston pin, while other designs will use dual spiro locks on each side to ensure an even more fail-safe setup.

Now that one spiro lock is installed in each piston, take that cylinder's connecting rod and a piston pin and smear some clean engine oil or assembly lube in the connecting-rod small end bushing and over the piston pin itself. Also lubricate the piston pin boss where the piston pin will ride. It doesn't need to be slathered on. Just add enough to get by until the engine starts; any oil is better than no oil. Slide the piston pin into the side lacking the spiro lock just enough so that it moves easily. If you push it in too far you won't be able to install the rod. Align the connecting rod to its proper orientation, line up the rod small end to the piston pin bore, and finish installing the piston pin by sliding it through the

The first order of business is to install one piston pin retaining wire (spiro lock). Installing the first lock without the connecting rod or piston pin in place is easiest. You can slightly stretch and expand the coils of the spiro lock to aid in installation and once an edge is started, work your way around the spiro lock until it is fully seated.

connecting rod and into the other side of the piston pin boss until it seats against the previously installed spiro lock. Finish off the assembly by installing the remaining spiro lock. If you thought the first spiro lock was hard to install, this one is a little harder fight to get right, as you now have the piston pin and connecting rod taking up some room. Using a rod vice or a regular bench vice with the protection of a shop towel, you can mount the connecting rod and stabilize the piston to help install the clips. Once you get to the last spiro lock, you'll get the tricks down to installing these.

Align the piston and rod in proper orientation, with the connecting rod bearing tangs faced away from the valve relief location, or on opposite sides of the piston centerline. The bearing tangs face away from the center of the engine when the piston is installed. Coat all friction surfaces with oil and slide the piston pin into place.

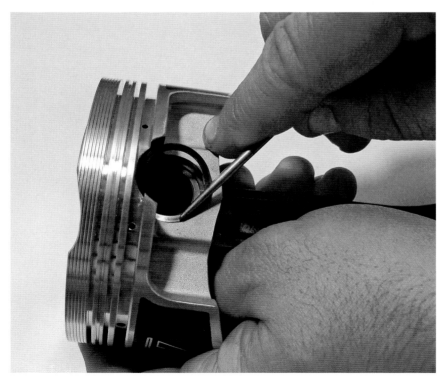

Much like the first spiro lock, stretch the lock coils out to expand the gaps a tad. Once the first coil end is installed, spiral the remaining length into place using a non-metallic object. (That scribe is plastic. I swear!)

PISTON RINGS AND GAPS

Up next is the sometimes tedious job of file-fitting rings for the cylinder bores. Piston rings are normally 0.005 inch oversized for the intended bore size. The 4.000-inch bore pistons will come with 4.005-inch piston rings that won't physically fit without some sort of cutting to open the rings' end gaps. The good news is that only the two compression rings need attention. The oil rings are normally good to go as supplied. Just measure to verify this before assuming all is right. For optimum results when file-fitting, you need to know the application.

Naturally aspirated engines have different ring gap specifications than a power-adder setup would have. Naturally aspirated engines will always have slightly tighter ring gaps than power-adder engines. You want to leave enough gap so that when the engine is under the harshest conditions, the ring gaps do *not* touch. When the ring gaps touch, it puts a huge amount of friction on the cylinder walls in the form of forced ring drag. This leads to cylinder wall scuffing and can tear up pistons and cylinder walls, leading to a full engine rebuild. The larger the bore size, the larger the ring gap will be by application. The ring gaps are normally calculated by a percentage of engine bore size. Typically, you will take bore size and multiply it by a set amount of gap per inch of bore size. If we use 0.006 inch per 1 inch of bore for a nitrous engine, apply it to a 4.030-inch bore (4.030x0.006-inch ring gap) = 0.024 inch of desired ring gap.

Piston and ring manufacturers have general guidelines for recommended minimum ring gap specifications by application. Final ring gap sizing decisions are left in the hand of the engine builder. There are many viewpoints and ring gap theories, and if you ask five machinists what they do, you will likely have five different answers. One thing that benefits LS-series engines is to keep the second ring gap a tad larger than the top ring gap. Being that they are "compression" rings, this goes against many ideologies that hold that the role of the second ring is to capture the blow by from the top ring. In

General Piston Ring End Gaps by Application

Top Ring Gap

Bore Size	N/A Street/Race	Mild P/A	N/A Race Only	Race Boost	Race N20
Bore X	0.0045	0.0050	0.0055	0.0065	0.0070
3.905	0.018	0.020	0.022	0.025	0.027
4.000	0.018	0.020	0.022	0.026	0.028
4.030	0.018	0.020	0.022	0.026	0.028
4.060	0.018	0.020	0.022	0.027	0.029
4.085	0.019	0.021	0.023	0.027	0.029
4.100	0.019	0.021	0.023	0.027	0.029
4.125	0.019	0.021	0.023	0.027	0.029
4.130	0.019	0.021	0.023	0.027	0.029

2nd Ring Gap

Bore Size	N/A Street/Race	Mild P/A	N/A Race Only	Race Boost	Race N20
Bore X	0.0055	0.0055	0.0060	0.0070	0.0075
3.905	0.022	0.022	0.024	0.027	0.029
4.000	0.022	0.022	0.024	0.028	0.030
4.030	0.022	0.022	0.024	0.028	0.030
4.060	0.022	0.022	0.024	0.029	0.031
4.085	0.023	0.023	0.025	0.029	0.031
4.100	0.023	0.023	0.025	0.029	0.031
4.125	0.023	0.023	0.025	0.029	0.031
4.130	0.023	0.023	0.025	0.029	0.031

fact, the top ring does 99 percent of the work of keeping blow by in check, while the second "compression" ring is more of an oil ring than anything.

The problem with having the second ring smaller than the top ring gap is that combustion will always be captured in between the two compression rings with nowhere to go. This usually leads to second ring flutter when that captured combustion attempts to escape in any way it can. The top ring cannot seal all combustion. Whatever amount of combustion makes it past the top ring, you do not want in between the rings. A larger second ring gap helps alleviate ring flutter issues. If the second ring flutters, this will lead to increased oil consumption as the second ring cannot scrape residual oil off the cylinder walls that is left behind by the oil rings.

Now comes the tedious task of file-fitting the piston rings. Using a ring square or a donor piston, carefully insert the ring you are measuring into the cylinder and square it up to the deck surface.

This particular second ring had less than zero ring gap so before we can measure the gap we need to put a gap there. When you are in situations like this, there will be quite a bit of filing needed on every ring used and it would not hurt to grind each ring a base amount from which to measure the gap. Something like 0.010 inch gap is enough.

The tools you will need to ring file your pistons are a feeler gauge set in the 0.016- to 0.028-inch range, a ring squaring tool in the bore size of your engine or a similar bore-sized piston used carefully, and the ring filer itself. When file-fitting piston rings, you can use a manual ring filer or an electric ring filer. They both effectively do the same thing; one is just much quicker than the other. If you're going to do more than one engine a year, use an electric filer. If not, use the manual ring filer.

The most important thing other than ring gap size is making sure the ring ends are square when cutting. This ensures that the gap is the same from the cylinder wall edge of the ring through to the inner diameter of the ring when in the cylinder. You need to measure the ring gap 1 inch down from the top of the cylinder. Use the ring squaring tool to push down the ring, and then measure the gap with your feeler gauges. When the gap is at your desired size, the rings will hold the feeler gauge in place, but you won't have to force the feeler gauge into the ring gap. It will slide in easily but have just enough drag to hold the feeler gauge by itself. I like to do all top compression

Again, we have absolutely no gap, so we need to file fit a base amount from which to start. We don't know if this is 0.000 or -0.010 at this point. Take off 0.005 to 0.010 inch at a time until you are in the positive gap range to have a good reference point for proper ring filing.

Ring filing is a tedious but necessary part of short-block assembly. An electric ring filer such as this one from Pro-Form will make for quicker ring filing. This ring filer has a dial indicator to measure how much you are cutting, but beware: 0.001 inch on this indicator does not equal 0.001 inch when in the cylinder. Use caution and make small cuts until you get to your desired gap.

rings first and then all of the second rings as a group, so I don't have to change technique for every other ring.

Once you finish fitting a ring and have your desired ring gap, you need to deburr the edges of the cut you performed to the rings so there are no sharp edges to cut into the cylinder walls or the piston itself. To deburr involves removing the displaced material caused by removing the sharp edges left from filing the ring gaps, by using a small file, or carefully bumping the corners of the cut surface against the ring filer to alleviate sharp edges.

The finished ring may be kept in its matching bore location until installation onto its mating piston, or you can remove it and mark its cylinder location number. If you are keeping them separate from the block, you can attach a piece of masking tape with the cylinder number on it or lay the rings out in order on a large cardboard sheet with corresponding cylinder numbers marked. If you end up mixing up the rings, this is not a huge problem, but you will have to go back and measure all rings and possibly move them from cylinder to cylinder to find the closest match.

Ring filing is take take take, until you achieve proper size by cylinder. You cannot just cut 0.020 inch off one end and have a 0.020-inch ring gap. It doesn't work that way. Using feeler gauges, check ring gaps often until you get to your size. Do this for all oil, second, and top compression rings.

INSTALL RINGS ONTO PISTONS

If building a stroker LSX, the pistons likely came machined for and equipped with an oil ring support rail. This does not need to be gapped because it only fits a certain range of bore sizes. The oil support rail helps to add structure under the oil rings where the piston was cut for the piston pin according to how long your stroke is. Your piston pin will need to be moved upward by half of the additional stroke and often is cut into the lower support of the oil ring land itself. You need to install this next. It will have a raised dimple to prevent it from

rotating on the piston when the rings are installed. To make it easy, just install the support rail with the dimple facing the "F" or the dot signifying the front of the piston. It can go either way, but uniform procedures like this throughout the engine build help keep ring alignment consistent in the following few steps.

Once the support rail is installed on all pistons, the next ring that may be installed is the oil ring expander. This is the wavy-looking ring that pushes the oil ring scrapers against the cylinder wall. Align the oil ring expander gap the same way as

This ring is the support rail necessary for large-stroke engines with a shorter piston compression height. The piston pin machining protrudes into the oil ring location, so a support rail is necessary to hold the oil rings in place.

The oil ring's wavy expander ring comes next. Align the expander ring gap to the front of the piston. Then install each oil ring scraper; top, then bottom is easiest. Align the gaps 90 degrees from each other. If you think of the piston as a clock with the gap in the expander ring at 6 o'clock, the two oil ring gaps would be at 10 and 2.

the support rail, toward the front of the piston. Next, install the upper oil ring scraper and then the lower oil ring scraper with at least 90 degrees between each oil ring gap.

Now install the second compression ring. Don't expand it so much that the ring breaks. An inexpensive piston ring expander would help the novice here, but most veteran engine builders just use their thumbs to expand the gap, while using the rest of their fingers to guide the ring over and down the piston to its proper groove. Do not twist the ring into place. If it gets hung up or stuck, remove it and try again, taking care not to break the ring in half. Once in place, make sure the ring rotates easily around the piston and sits well into the ring groove.

Finally, install the top piston ring. This installs in the top groove of the piston, in a similar procedure as the second compression ring. Once both compression rings are in place, align the gaps to their proper orientation, 180 degrees apart from each other and offset from the oil scraper ring end gaps. You'll have to check this alignment once more before installing the piston, ring, and rod assembly into the cylinder bore.

Next up is the second compression ring. It is safest to install these with ring expander pliers, but once you are comfortable with installing rings, you will get used to just using your thumbs to expand the ring and your forefingers to guide it over the piston and into place.

The final piston ring to be installed is the top compression ring. Installation is like the second compression ring. When both rings are installed, align the second ring gap 180 degrees from the center of the two oil ring gaps and align the top ring 180 degrees from the second ring gap.

MEASURE ROD BEARING CLEARANCE

This step can be done at any point prior to installing the piston, ring, and rod assembly into the cylinders, even before any block machining has been completed. Ideally, this should be done before the pistons and rods are mated. To cover all bases, every engine oil clearance should be measured. The connecting-rod bearing oil clearance is no exception to this rule. At this stage you more than likely already checked at least the main bearing clearance and have an idea of how to go about checking oil clearances. Rod bearings are not much different from main bearings as far as measurement technique, with the same procedures and measuring tools used. You need to know the actual crankshaft rod pin (journal) diameters measured with a micrometer.

Set your dial bore gauge to the identical size of the rod pin and measure the two mating connecting rods for that rod pin. Mating cylinders are 1–2, 3–4, 5–6, and 7–8. It is highly possible that you'll only need one measurement and dial-bore gauge setting when using all-new parts, as the crankshaft pins will be very close to each other. Manufacturers such as Callies Performance Products keep tolerances within 0.0001 inch from one rod pin to the next. The bottom line is that these components still need to be verified by measuring and then documenting these readings.

Once the connecting rods, pistons, rings, and rod bearings are in place, the obvious next step is to install these piston and rod assemblies into the block. Separate the connecting-rod caps from the connecting rods. A rod vice is extremely helpful here, as most aftermarket connecting rods have steel alignment dowels that are a tight fit. If the rod bearings are not installed, you may install them in preparation for short-block assembly. If you need chamfered rod bearings for aftermarket crankshafts, these have an upper and lower location similar to main. Only one side of the rod and bearing need the chamfer. The rod bearings marked "upper" are intended for the connecting-rod side. The "lower" rod bearings are for the bearing cap. The bearings chamfer should line up to the connecting-rod chamfer itself and form a uniform taper along both components. Stock rod bearings intended for the factory crankshaft have no upper or lower. Either bearing insert is correct. Line up the bearing tabs and install them. Keep the rods and matching rod caps together. They cannot be swapped with other rods. Most engine builders will scribe the cylinder numbers on both the rods and rod caps so that they are not installed in the wrong location.

If using aftermarket connecting rods/crankshaft, either chamfered rod bearings or narrowed rod bearings must be used. Because these are offset bearings, there is an upper (rod) and lower (rod cap) location for each bearing half.

Install the upper bearing half into the connecting rod and the lower bearing insert into the connecting rod cap.

With all eight rods and rod caps separated and all subparts installed, you can now prepare the block. It should be fairly clean if you used an engine bag to store the block while working with the pistons, rings, and rods. When building engines, it becomes normal practice to wipe the bores with a clean lint-free cloth, and it's good form to also clean the crankshaft rod journals. Now that all the cylinders and crankshaft journals are cleaned, we may finally progress to installing the pistons. You need some sort of piston ring compressor to collapse the rings into cylinder bore size. There are two basic types of ring compressors. First is the band type, which is a steel band with an expandable gap. The gap side will have either a cantilever setup or a place to install a matching pair of ring compressor pliers. The other installation setup, which is simpler to use and easier on rings, is a tapered ring compressor. There are a few manufacturers of tapered ring compressors in a variety of bore sizes. A tapered ring compressor should work as long as it is within 0.005 inch of the actual bore size.

Using engine assembly lube or clean engine oil, lube the piston rings, connecting rod bearing, and cylinder walls in preparation for further assembly.

If using a conventional bore size, you can use tapered ring compressors for quick ring compression. Tapered ring compressors gently compress the piston rings as the piston travels through the compressor until the piston is in the cylinder.

The pistons can be installed in any order, but it's easiest to install one entire bank in a group, then the other bank afterward. I always start with cylinder no. 1 and end with no. 8, simply as that's what I've always done. There is no set order as long as progress is made. In preparation for piston installation, spread clean engine oil onto the piston rings, cylinder walls, and pistons. Prep the crankshaft rod journals by smearing on engine assembly lube. You may also apply some engine assembly lube to the rod bearings themselves. If using ARP hardware, which is very common in aftermarket connecting rods, now would be a good time to apply ARP lube or extreme pressure lube to the rod bolt threads and under the washer-head of the bolt. The proper thread lubricant is specified by the bolt or rod manufacturer.

Finally, it's time to install the piston and rod assembly. Choose a cylinder to start with (no. 1 in my case), and rotate the crankshaft until the chosen journal is at bottom dead center for that cylinder. Obtain the piston ring compressor, making sure to lube the tapered portion of the compressor before each piston and rod assembly is installed. I've found it easiest to put the ring compressor on a flat surface and lower the connecting rod into the ring compressor itself, then pull the compressor up from the bottom over the piston, compressing the rings as it gradually travels upward. Don't go too far, just enough to hold the piston in place without the compressor moving. Guide the piston and rod assembly into the cylinder and gently lower it into place. Pay particular attention to the piston orientation ("F" toward front) and connecting-rod versus crankshaft location. You don't want to nick the crank journal.

It is best to have the corresponding crankshaft rod pin close to BDC for installation access of the rod cap itself. Finish lowering the assembly until the bottom of the compressor rests on the cylinder deck surface. Using a piston hammer (rubber dead blow hammer) or the wooden end of a hammer, gently tap the top of the piston until it is in the bore. You will get a feel for what is right and when there are problems. You shouldn't have to tap the piston hard for it to compress the rings, but if the piston doesn't budge, don't force it. Check the rings for proper fit and movement. If no problems are found, attempt to reinstall it.

When the piston is in the cylinder and the ring compressor is free, the piston can be pushed by hand or by using a soft object the rest of the way down into the cylinder. Guide the connecting-rod end over the crankshaft journal. Sometimes the rod bearing will fall out or be displaced by the tapping required to install the piston. If this happens, you can usually just slide it back into place and go on. If the piston and rod assembly is now in place at bottom dead center with no issues, you can now install the rod cap. Do not torque the rod bolts at this time though; wait until all pistons are installed. Push the bolts down enough to fully seat the rod cap.

If you are using a unique bore size, you may need to use a band-type ring compressor. These are likewise easy to use, but they take some adjusting to get perfect. Compress the rings and gently tap the piston into place using a rubber hammer.

When the piston is in the cylinder, push it farther down while guiding the connecting rod bearing over the crankshaft rod pin.

The first bank of the installation will go easy once you have a feel for it. The opposite bank of cylinders is also not too bad until the connecting rods are close to being installed onto the rod pins. At this point, there is already a connecting rod on each rod journal. You may have to push the opposing cylinder's rod forward or back to create more side clearance on the rod journal. With the second bank rods, you also have to guide the rods more carefully, as the rod has to be lined up really well to its journal on both sides now. With the first bank's installation, you didn't have another connecting rod installed to fight against for clearance. Install all other piston and rod assemblies in sequence until finished. Be careful not to nick the rod bearings, cylinders, pistons, or journals. If you do, and you can feel the damage with your fingernail, it will need to be repaired. Usually you can remove the crankshaft and have it polished.

Use ARP moly lube on the ARP connecting-rod cap screws and under the washer face of the bolts. Apply a film of assembly lube to the lower rod bearing insert and install the matching rod cap. Then install the connecting-rod cap, aligning the bearing tangs together on the same side.

Tighten each cap screw equally to pull the rod cap and connecting-rod into place until seated. Do not torque individual rods until each rod pin has both connecting rods in place. Repeat this procedure for all eight piston and connecting rod assemblies.

When all eight piston and rod assemblies are installed, turn the entire engine over on the engine stand to gain access to the rod bolts. They should not be torqued yet, but using a quality torque wrench, tighten all rods to a generic setting such as 30 lb-ft and verify that there are no interference issues with the rotating assembly. Remember, all moving parts need a minimum of 0.050-inch clearance between themselves and other moving parts on nonfriction areas. These critical areas are piston skirts to crank counterweights, connecting-rod bolts to bottoms of cylinder sleeves, and a no. 8 piston to crankshaft reluctor wheel (pressed onto the rear of the crankshaft).

Now is also a good time to check rod side clearance. You can use thin feeler gauges or a dial indicator to check the clearance and document your findings. Rod side clearance problems are not common and usually would err to the tight side when wrong, although it's better to be safe than sorry and measure. You can use a dial indicator affixed to the engine block to measure the side clearance of each rod pair or with a feeler gauge set. Too tight of clearance would cause friction and binding when the connecting rod heats up. The rod side surface (or crankshaft journal sides) would need to be ground at a machine shop for adequate clearance if too tight. The OEM specification is 0.005 inch to 0.020 inch, although for performance builds, 0.015 inch to 0.025 inch is adequate.

If everything checks out well, we can move on to torqueing the connecting-rod bolts. If you didn't measure rod bearing clearance previously with the dial-bore gauge and micrometer, you may now check it with the Plastigage. Torqueing the rod bolts is straightforward; you torque the rod bolts in pairs on each connecting rod. Refer to the manufacturers' connecting-rod specifications for bolt

Once the pistons and connecting rods are installed and the rod caps are initially seated, we can check the rod side clearance using a dial indicator.

torque specs and bolt lubricant requirements. With new connecting rods, it's recommended to retorque the bolts several times to ensure they are tightened to full torque values. With all aftermarket rod bolts, there is also a bolt stretch measurement specified. This measurement is the preferred method of determining bolt torque amounts. The rod bolts stretch different amounts, depending on bolt material, length, and diameter. Refer to the manufacturer's requirements on bolt stretch specifications. Bolt stretch is a little harder to measure on LS-series engines, but a few tool companies make specific rod bolt stretch gauges to confirm proper bolt torque values and stretch amount.

If reusing stock rod bolts, there is a certain angle procedure for torqueing these bolts, but given the reasonable price of upgraded ARP bolts, you really shouldn't reuse stock bolts. Keep in mind that any time you change the fastener on any engine part, it needs to be resized to be right. You may be able to get away without resizing if you check the connecting-rod big end for distortion. Stock pre-2001 rod bolts are supposedly weaker than the 2001+ connecting-rod bolts. The torque specification is different between the model years, with 2001 being the change year.

Once the rod bolts are torqued properly, you now have yourself an assembled short-block. You can go many directions from here; you can move on through the next chapter and build the long-block, or you can set it aside in a sealed engine bag for a few months to build at a later time if desired. If after reading this section, you don't feel you want to build your short-block, or you don't feel comfortable with engines or the precision requirements of setting up a short-block, you should hire an engine builder to build an engine to your specifications.

When all connecting rods are in place, move to the torqueing procedure. There are two ways to do this: by following a torque specification value or by bolt stretch. When torqueing to a set value, loosen and tighten each connecting-rod bolt at least three times to fully seat it into place.

I prefer to tighten with a torque wrench and verify with a stretch gauge. You need to zero the stretch gauge to each individual rod bolt measurement. Using bolt stretch is more accurate versus friction torque measuring. The rod bolts I tightened to torque value in this engine ended up with exactly the correct bolt stretch.

Chapter 7
Long-Block Assembly and Inspection

INTRODUCTION TO LONG-BLOCK ASSEMBLY

You may wonder why I chose to separate information on short-block and long-block assembly into two chapters when the engine assembly is considered by some to be a single topic. One reason is that we are covering a plethora of information for each assembly procedure. The other is that the finished short-block is a good halfway point in the entire engine assembly procedure, a "Y" in this road that can head (no pun intended) in any direction. The techniques and assembly change at this point; you are not directly dealing with oiling or bearing clearances anymore. Rather, we are dealing with the top-end valvetrain-related components that in turn relate to the performance tendencies of the engine. While the short-block is responsible for accommodating the stresses of rotation and engine loads, the components that finish the long-block are the cause of these concerns.

LS-series engines have the versatility to be used in a variety of applications. This wet-sump 427-cubic-inch LS7 is destined to be in a first-generation Camaro, which is an easy and popular retrofit. An engine package like this would typically make horsepower in the 650 to 700 range.

In this chapter, we will discuss the installation of camshafts, lifters, timing chains, cylinder heads, and remaining valvetrain components. Procedures for degreeing the camshaft and the effects of changing camshaft timing will be covered. The procedures for measuring piston-to-valve clearance, checking and adjusting lifter preload for hydraulic camshafts, and setting valve lash for solid-roller lifter camshafts are all included. After following the installation procedures of this chapter, you will have a completed long-block LS-series engine assembly, ready for all external covers and, shortly afterward, installation into your vehicle.

Once the short-block assembly is finished, we move on to what is called the long-block assembly, which basically adds the heads and camshaft components. Start by installing the crankshaft keyway (if not pre-installed) using a brass punch and hammer. The keyway indexes the crankshaft gear, aligning it with TDC on cylinder no. 1 of the rotating assembly.

CAMSHAFT AND TIMING CHAIN

With any luck, your short-block is still on an engine stand and ready to progress further. You can now take your camshaft of choice and prepare it for installation. You do not want to just slap it in and forget about it; camshafts are full of debris from the machining processes and from packaging. You want to clean the camshaft thoroughly, preferably with cleaning solvents or some sort of aerosol cleaner that leaves no residue, such as a quality brake parts cleaner. Once cleaned of machining oils, grease, and debris, you should carefully and lightly use compressed air to dry it and make sure absolutely nothing is left over from cleaning. Once this is done you may lube it with the engine assembly lube of your choice, except for the lobes

The camshaft is ready for installation after first verifying specs, cleaning the machining debris and rust prevention coating, and then generously coating it with engine assembly lube.

Using a camshaft holder or a long bolt or threaded rod, slowly and carefully install the camshaft into place. It should not bind or lock up while installing and should not require brute force to install.

you will be degreeing and checking PtoV clearances with. These two lobes are the first and third lobe from the front of the camshaft; the first lobe is the cylinder no. 1 intake, and the third lobe is exhaust for cylinder no. 1. Before the final install of the actual lifters used, you may lube these two lobes, but for accuracy during inspection and measuring, it's best to leave oils and assembly lubes off of that portion of the camshaft.

If not measured previously, now is a good time to measure the cam journal oil clearance, before cam installation. This is not something many builders check, and it's not specifically required unless you suspect an issue. This procedure is similar to measuring main journal clearance, although little can be done to change camshaft journal oil clearance. You will need a dial-bore gauge and a micrometer. First, measure the corresponding camshaft journal and then zero the dial-bore gauge to that measurement. Using the dial-bore gauge in the matching cam bearing location, a positive number indicates oil bearing clearance, while a negative number indicates a problem and should result in a camshaft that cannot be installed or locks down if installed. The Plastigage cannot be used to check cam journal oil clearance. Don't try it.

With the camshaft coated and lubed, you may now carefully slide the camshaft into place from the front of the block to the rear. The use of a camshaft holder to help installation is recommended but not required; a makeshift cam holder can be created easily with a long bolt. An OEM M12 crankshaft bolt will do. The main focus here is to slide the camshaft into place without nicking and damaging the camshaft bearings or knocking them out of place. The camshaft should easily slide into place if the cam bearings were installed correctly and then tested with the "test" camshaft. Some light resistance is normal, but if the camshaft feels like it hits a wall halfway through, then further inspection is required. Do not force the camshaft into place under any circumstances.

LS-series engines use a camshaft thrust plate to keep the camshaft from walking forward and back. Upon initial installation, such as in a new engine build, use assembly lube on both sides of the thrust plate to prevent wear until engine oil takes over.

Once the camshaft is in place, we can install the camshaft thrust plate. This is a dual-purpose piece. The obvious function is to hold the camshaft in place as it rotates and prevent camshaft walk forward and rearward. Its other function is to seal the front of the lifter galley oiling holes. If the camshaft thrust plate is loose, or the orange captured O-ring seal is missing or damaged, a loss of oil pressure will result. I prefer to use new camshaft thrust plates with all new engines, but using a donor plate from your old engine is fine also. Look it over for any wear or problem areas before getting this far with your build. If there is any question as to its condition, replace it. Using Loctite on the four (six on GMPP LSX) thrust plate bolts, torque these to 18 lb-ft using conventional bolts, or if using countersunk bolts, torque these to 11 lb-ft. Excessive torque with countersunk bolts will crack the camshaft thrust plate.

The newer camshaft thrust plates use four countersunk M8 bolts with Torx heads instead of the more common hex-head bolts. When using hex-head bolts, make sure the bottom bolt is the correct length as it protrudes into the crankcase and can interfere with the rotating assembly.

Torque the countersunk T-40 Torx bolts to 11 lb-ft using Loctite on the threads if reused. Torque hex-head bolts or ARP bolts to 18 lb-ft.

CHECKING CAMSHAFT THRUST

This measurement is often overlooked, but it warrants inspection just like any other engine friction surface. We are checking for zero-or-less thrust clearance or excessive thrust clearance. Zero clearance will bind the camshaft and timing gear and effectively clamp or sandwich the thrust plate, causing direct metal-on-metal wear. In a worst-case scenario this leads to shearing oil shavings into the oil and the thrust plate possibly cracking. Neither is a good situation to have on a fresh engine. Too much clearance can also affect the camshaft sensor signal. Really excessive clearance can accelerate timing chain wear, although too much clearance won't be the case normally. Clearance specification for camshaft thrust is 0.001 to 0.012 inch. Typical clearance tends to reside on the high end of the scale.

There are a few ways to eliminate problems here before they happen. You can create a mock-up of the assembled camshaft, thrust plate, and upper cam gear on a bench before installing it into the engine and use a feeler gauge set to measure the clearance. The easier and preferred way, however, is to check after installing the camshaft thrust plate and upper timing gear. Use a dial indicator in a similar fashion as when checking the crankshaft thrust clearance. Using the dial indicator for measuring thrust, move the camshaft forward and back and note the travel amount. Compare to specifications and document the measurement.

TIMING CHAIN

After making peace with the camshaft thrust clearance, we can install the complete timing chain set. There are quite a few manufacturers of these, but I'll show the installation of a stock timing set and then one popular aftermarket

If the camshaft does not turn when the timing gear is torqued down, this could be due to improper depth camshaft snout machining. To check this clearance before installing, you can mock-assemble the camshaft, thrust plate, and timing gear. The thrust plate should spin freely; further, you can measure the clearance with a feeler gauge.

set. With the stock set, it's pretty straightforward: install and go. Many aftermarket pieces have multiple crankshaft keyways for advancing and retarding the camshaft timing. To start off, find the zero advance markings and install the lower timing set gear to the zero advance position as a baseline setting. Note that the crankshaft keyway position while cylinder no. 1 is at TDC is not at 12 o'clock. Rather, it's at the 1:30 position. It points 45 degrees from the crankshaft gear's timing mark or, more easy to remember, to the no. 1 cylinder location itself.

Many aftermarket timing chain sets are adjustable by keyway slots in the crankshaft gear. Remember that when you change the camshaft timing using the crankshaft gear, you need to use the new corresponding timing alignment mark for each change: 0 to 0 and +2 to +2, and so on.

LS-series engine oil pumps are driven by the crankshaft gear. Some timing gears are two pieces, while OEM and few other brands are one piece. If two pieces, remember to install the oil pump drive gear at some point during your engine-building task.

Install the lower timing gear in its baseline setting, making sure the crankshaft keyway is in place first. The LS crankshaft timing gear, in addition to driving the timing chain, also functions as the oil pump drive gear. The stock gear integrates both into one piece, and so do many aftermarket gears. Some recent aftermarket manufacturers have been marketing two-piece crankshaft gears, with the timing chain portion separate from the oil pump drive gear. These are usually a little easier to both install and remove, so camshaft timing changes are slightly easier to make. The one-piece integral gears usually need a three-jaw puller to remove once pressed into place. Keep this in mind if any cam timing changes need to be made and be prepared with the proper tools for cam gear removal. Many engine builders will "press" the crankshaft gear into place using a BFH (hammer), but it's recommended to use a long threaded rod or a long bolt and a spacer to press the gear into place if available. Even using an old stock timing gear flipped backwards with a stock crankshaft center bolt is a more effective procedure than employing a hammer and a few grunts.

Once the lower gear is pressed into place, verify TDC and install the upper timing gear and timing chain as a unit. You may need to rotate the camshaft or crankshaft slightly to line it up into the right location. Line up the camshaft timing dots. The camshaft gear's mark should be at 6 o'clock, while the crankshaft timing mark should be at 12 o'clock. Tighten the camshaft bolts to 26 lb-ft using Loctite on the threads.

DEGREEING CAMSHAFT

Degreeing the camshaft is an important process to do once internal engine work is done. There are a few procedures for checking and verifying camshaft location, but we'll cover the short-block version with the heads off in this part of the text.

Anytime you change a crankshaft, timing set, or, more obviously, the camshaft, the camshaft and components will need to be degreed in. The process is useful more to verify proper timing than change it, as nine times out of ten, it will be correct as is. All it takes is that one time that you don't check for it to be wrong, and I assure you that you will never overlook camshaft timing again. The factors that may affect camshaft timing in an engine are the crankshaft keyway location, camshaft dowel location, lifter bore location, timing gear set slack and machining, and the camshaft manufacturer's machining itself. You are normally dealing with a variety of manufacturers, with each having its own tolerance levels for its own parts.

To degree the camshaft, you need a few specialized tools and some background knowledge on camshaft timing events. You will need some form of degree wheel, a degree wheel pointer, a piston stop or dial indicator on a bridge fixture to determine absolute TDC, and at least one additional dial indicator to measure camshaft lobe lift. You also will need a sample lifter. A solidly made converted hydraulic lifter is suitable. The lifter needs to ride on the intake lobe on the camshaft for the cylinder you are checking. It is easiest to check cylinder no. 1 due to ease of access. We also know that aligning the timing chain marks gets us close to TDC for cylinder no. 1.

While some gear sets will just slide into place with hand pressure only, the vast majority require pressing into place. To do this you will need a spacer (a flipped old gear), an M16 threaded rod or bolt, and a few washers.

Install the timing set while lining up the camshaft alignment mark to the corresponding crankshaft mark at TDC. Sometimes the camshaft alignment dowel is too large and will bind in the upper timing gear. If so, the timing gear dowel hole will need to be honed ever so slightly to fit.

Using premium fasteners with a dab of Loctite, start the three camshaft bolts into place. They do not need to be torqued while degreeing the camshaft, but it is good practice to do so. OEM bolts are torqued to 26 lb-ft, and ARP bolts are torqued to 28 lb-ft.

To start the degreeing procedure, two lifters must be installed. These can be old lifters or homemade solid lifters, such as those used in the piston-to-valve (PtoV) section later on.

First you should have the camshaft and timing set installed into your short-block. Install your modified lifter onto the camshaft for intake lobe no. 1, which happens to be the first lifter at the front of the engine on the left bank (driver's side). If using the factory plastic lifter guides to hold the lifter in place, note that the plastic guides fit tightly over the lifter and are likely to skew your results. I recommend installing the lifter 90 degrees from the camshaft centerline as it would ride in a running engine, although without the lifter guide. We're not rotating this engine very fast during these measurements, so just make sure the lifter roller is always parallel to the camshaft rotation.

Once on the opening and closing ramps of the lobe, the lifter will align itself, and the dial indicator may help to keep it aligned also. If you're using an iron block, you may use a magnetic base dial indicator. If you've got an aluminum block, you will need to bolt on a steel plate to the deck or use a long bolt as the fixture for the dial indicator. You will need to use a dial indicator extension so that you can position the indicator where you can read the numbers easily. Align the dial indicator rod with the same orientation as the pushrod's travel for the most accuracy. Zero the dial indicator with the lifter positioned on the base circle of the camshaft. With LS-series engines, this is where both timing set alignment marks are at 12 o'clock for cylinder no. 1.

Degreeing the camshaft aligns the camshaft events with the crankshaft events, so we need to know where each of these is in relation to the other. We do this by using a degree wheel and dial indicators.

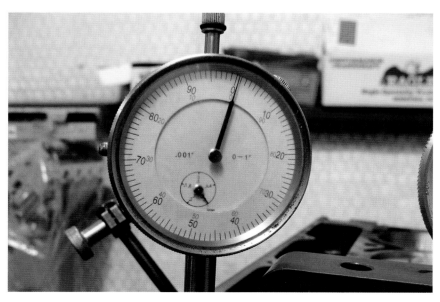

First zero the camshaft lift dial indicator while the lifter is on the base circle. The dial indicator is measuring the no. 1 intake lobe lift at the lifter installed previously. The dial indicator measuring the piston location needs to also be zeroed out at TDC.

It's helpful to turn the engine 45 degrees toward the passenger side on the engine stand so that the cylinder bank you are working on is completely vertical. This helps while using the bridge and keeps everything relatively flat and easier to read, as if you were working on a flat table.

Now align the timing chain gear marks. Install your degree wheel, keeping note to align TDC on the degree wheel in close proximity to a bolt hole that your degree pointer may mount from. You will need to set actual TDC more accurately in the next step, but this is close enough to start as a baseline adjustment. I'll call this "rough TDC." Accuracy here is in the +/- 4 degree range due to piston dwell time at TDC and not yet close enough to accurately degree the camshaft. To find absolute no. 1 piston TDC, you need to use either a piston stop bolted to two adjacent cylinder head bolt locations across the cylinder, or by using a bridge and dial indicator. Each method is a different procedure but essentially will show the same end result, which is finding absolute TDC. I prefer the dial indicator method myself as it also tells you how far in or out of the block the top of the piston is.

To find absolute TDC using a bridge, you can leave the piston at rough TDC, adjusting the dial indicator to show "0" and making sure it is preloaded enough against the piston so that it follows the piston down the bore as it reaches and leaves TDC. Using a dial indicator on a piston is most accurate in the dead center of the piston. If that is not possible due to piston shape, you can deviate along the piston pin length front-to-rear as necessary. You don't want to measure from the top or bottom of the piston crown, as readings will be skewed from "piston rock" on a cold engine.

With the dial indicator zeroed out at rough TDC, turn the crankshaft over clockwise until you reach 0.100 inch of travel and note the degree wheel reading or mark the degree wheel with a dry erase marker. Now continue rotating the engine in either direction until the no. 1 piston is getting close to TDC. You want to stop at that same 0.100-inch reading as the piston is traveling upward and note or mark the degree wheel reading. Exactly halfway between these two measurements on the degree wheel is absolute TDC. If one number is 15 and the other is 10, then TDC is one half the difference between or the average of the two numbers ((15-10)/2 = 2.5 degrees). In this case, you'd move or bend your pointer 2.5 degrees toward your other measurement, depending on which one is higher or lower.

This may take two or three tries to get exactly right. It doesn't hurt to verify these results a few times. Although unlikely, it's possible for both numbers to match. If this is the case, your pointer is already at absolute TDC and no adjustment is necessary. Once you determine absolute TDC and adjust your pointer to correspond, rotate the crankshaft a few times with the bridge still in place. As the dial indicator zeros out, your degree wheel should pass TDC. I prefer to leave the bridge in place while verifying camshaft location so that if I inadvertently bump or bend the degree wheel pointer, I have a backup reference point.

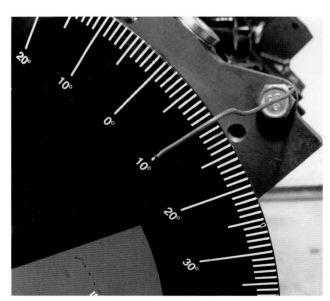

Install your degree wheel and a makeshift pointer as close to TDC as possible. You won't be able to get it perfect, but it will be close enough to tweak your pointer exactly in the next few steps. Rotate the crankshaft counterclockwise (CCW) until the dial indicator is at 0.100 inch on the piston, then rotate clockwise (CW) until 0.050 inch is shown. Note the degree wheel position. Ten degrees before top dead center (BTDC) is shown.

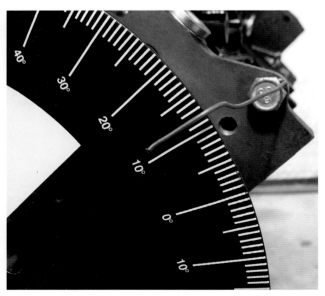

Next, rotate the engine CW until the dial indicator zeros out and keep rotating until we are at 0.050 inch again. This time we are at 12 degrees after top dead center (ATDC).

Now that we have our two degree readings, we need these two angles to match to find absolute TDC. We are not off much. To achieve the centerline we use a simple formula (12 − 10)/2= 1. Therefore, we need to move the needle 1 degree toward the smaller number, which is 11 degrees.

To double-check our work, repeat. In this case we would see 11 degrees at both 0.050 inch measurement settings, and of course halfway between these is exactly 0, or absolute TDC. When using a piston stop, these procedures for pointer adjustment are similar, but no piston dial indicator is required.

If you are using a piston stop, finding absolute TDC is similar to using a bridge, although you are not measuring piston travel. Rather, you are forcing the piston to stop before and after TDC with a fixed object and location. To use a piston stop, you have to first move the piston out of the way to facilitate bolting the piston stop to the block. You will need to use two opposing cylinder head bolts to place the piston stop in the center of the piston. If using stock head bolts, you probably will need some spacers or a stack of washers to lock down the piston stop against the deck surface, as you are not likely to find shorter M11 threaded bolts at your corner hardware store.

Once the piston stop is affixed to the block, you may lightly rotate the engine until the piston "stops" against the piston stop. Now you know where this tool gets its creative name. Once you have the piston against the piston stop, note the first degree wheel position the same as you would with a bridge. Then rotate the engine the opposite direction until the piston travels back up again and lightly against the piston stop from the other direction. Note this degree wheel position also. In the same fashion as finding absolute TDC using a bridge, take the higher degree number, subtract it from the lower number, and divide it in half. Then move the pointer this amount toward the zero position. If both degree numbers are exactly the same when rechecking, you've found absolute TDC. If it so happens that the numbers match before any adjustment is done (that is, 10 before and 10 after), then you are already at absolute TDC and no pointer adjustment is required. After the cam timing indicator is set, you may remove the piston stop.

Once true TDC is found and set on the degree wheel, you can move on to finding the location of the no. 1 intake lobe in relation to the no. 1 piston. The way camshaft manufacturers determine how much advance or retard a camshaft has is by the intake lobe centerline. On your camshaft card it may show

Now that we know absolute TDC is set, we can use the dial indicator to reference the camshaft location. Using the camshaft intake lift dial indicator, turn the crankshaft until peak lift is found. Rotate the crankshaft CCW until the dial indicator reads 0.050 inch down from peak lift and note the degree wheel number. Sixty degrees is shown here.

Rotate the crankshaft the opposite direction past peak lift until 0.050 inch down from peak lift is measured; 156.5 degrees is shown here. We take the first and second measurements at the same lift amount and add them together, then divide in half to find the intake centerline (ICL). The formula in this case would be 60 + 156.5 = 216.5 / 2 = 108.25 ICL.

a number for intake centerline. Let's say it shows 110 ICL, but it is on a 114 lobe separation angle (LSA). This indicates that the camshaft has 4 degrees of advance in it. There are a few ways to determine the location of the centerline of the intake lobe. One way is to set the lifter at peak lobe lift and zero the dial indicator at that point. Much like finding true TDC of the piston, you document the position of the degree wheel with 0.050 inch or 0.100 inch less than peak lift on the opening and closing ramps of the intake lobe. Using these two numbers, you can find the centerline between these two lift points. Take both readings, add them, and divide by 2 to find the intake centerline. If your camshaft is within 1 degree of what the cam card says for intake centerline, no adjustments are required.

The other method to find intake centerline (and the more accurate version) is by measuring the actual camshaft duration from a 0.050-inch opening to a 0.050-inch closing of the intake lobe, which are the camshaft duration numbers that most people relate to camshaft sizing. Using these numbers you can tell if the camshaft is the right size by duration and lift numbers, as well as the location of cam lift in relation to the piston. Remember that halfway between opening at 0.050-inch lift and the closing point at 0.050-inch lift is the intake centerline. With some degree wheels you may have to calculate this yourself, as some use 180 degrees before and after TDC instead of 360-degree increments. Ideally, your engine may not need any adjustments at this point for cam timing corrections, but what if you are off 2 or more degrees? After double-checking your work and verifying your readings, we can now progress to making adjustments if required.

CHANGING CAMSHAFT TIMING

If any adjustments are needed, there are a few options for changing or correcting camshaft timing. If you have a stock timing set, your hands are tied a little. Even then you can drill the upper timing gear and use offset camshaft dowel bushings to correct minor variances in camshaft timing. It is much easier when using an aftermarket timing set, though, as there are normally pre-marked advance and retard positions in the lower timing gear. If you are advanced 2 degrees too much, all you need to do is retard the timing gear 2 degrees. If you changed advance and retard using the lower crank gear, you have to be careful and verify the change's results. Depending on the manufacturer, you may move the cam timing more than desired. Most end up changing 1 camshaft degree per keyway position change, not the 2 as marked.

Be careful with keyway adjustments. If you advance or retard the camshaft with the crankshaft keyway position, you cannot use the same timing chain alignment dot you used for the zero advance position. If you advanced 2 degrees, you now have to use the 2-degree timing alignment mark. This is the same for any other setting. Once you deviate from 0, you cannot use that alignment mark any longer.

```
PART # 54-000-11    SN#: P 0553-08
ENGINE: CAMSHAFT, LS1 CHEVY LS1/GEN III '97-UP    PART#: 54-000-11
   TEXAS SPEED & PERFORMANCE LTD
GRIND#: LS1 254/260 112
SPC INSTR 1:    TEXAS SPEED
SPC INSTR 2:
                    INTAKE  EXHAUST
VALVE ADJUSTMENT    HYD       HYD
GROSS VALVE LIFT    .630     .625
DURATION @
 .006  TAPPET LIFT 304      312
VALVE TIMING   OPEN          CLOSE
@ .050   INT:    19  BTDC    55  ABDC
         EXH:    66  BBDC    14  ATDC
THESE SPECS ARE FOR CAM INSTALLED
@ 108.0  INTAKE CENTER LINE
             INTAKE   EXHAUST
DURATION @ .050  254    260           SPRINGS REQUIRED
LOBE LIFT       .3710  .3680   VALVE SPRING SPECS FURNISHED
LOBE SEPARATION 112.0          WITH SPRINGS
```

Referencing our camshaft card, we can find the desired intake centerline listed at 108 degrees. You will find that most camshafts are ground correctly and being that ours is 108.25 degrees, you are not going to get much closer than that. This also shows that the camshaft is ground with 4 degrees of advance.

Now that the camshaft is degreed and verified, we can torque the ARP camshaft gear bolts to 28 lb-ft with Loctite. If you torqued them previously, double- and triple-check these bolts for tightness at this time.

With any timing chain setting adjustments, you will want to go back and recheck all of your work by verifying camshaft timing once more before moving on with further assembly. There should be no reason *not* to check your work at this point. Once you are 100 percent sure of your camshaft settings, you may move on.

One last time: verify that the camshaft gear bolts are torqued to specification. This is especially important if you changed camshaft timing or left the bolts snug to take the measurements. For peace of mind, it wouldn't hurt to check before installing the timing cover as well.

CHECKING PISTON-TO-VALVE CLEARANCE

It is vital to check piston-to-valve clearance (PtoV) on all new engine setups and if changing camshafts or cylinder heads. If there is mechanical interference here, you will quickly do some major damage to valves and quite possibly a few pistons upon the initial startup or the first time the engine sees higher rpm. The crucial period where both valves get close to kissing the piston is during overlap, a period starting at 15 degrees before through 15 degrees after TDC. Notice that this is *not* where maximum lift occurs. That happens when the piston has traveled well into the cylinder, approximately 2 inches down on a 4-inch bore crankshaft on average. Maximum valve lift has little influence on piston-to-valve clearances.

The tools necessary to check PtoV are:

* Degree wheel
* Solid lifters or converted-to-solid hydraulic lifters
* Adjustable-length pushrod or adjustable rocker arms to set 0 lash
* 1-inch travel dial indicator or feeler gauge assortment
* Lightweight checking springs

If you don't have access to solid lifters, you can easily make your own set with some old donor lifters. You will need two hydraulic

lifters that you don't plan to reuse. Remove the c-clip at the top and set it aside. Now remove the innards of the lifter assembly. First is the pushrod cup, which will be reused, then a thin, flat washer, followed by the lifter plunger itself. Under the lifter plunger is a small spring. Discard the spring, as it's not needed. Reinstall the lifter plunger, pushing it down with a screwdriver until it bottoms out. Next you will need a flat 5/16-inch washer that fits inside of the lifter internal bore. The dimensions needed are 0.620-inch outside diameter and 0.055-inch thickness. You may need one or two flat washers to achieve the shim thickness. With the washer in place, test-fit the pushrod cup to see if the C-clip groove is accessible. You want just enough washers or shims so that you can reinstall the pushrod cup and c-clip back into place.

While you do not need a solid lifter to degree the camshaft, you do need one to accurately check piston-to-valve clearance. You can convert a donor hydraulic lifter to a solid lifter by disassembling it and either shimming it or filling it with epoxy. Remember you ideally need two lifters, one each for intake and exhaust.

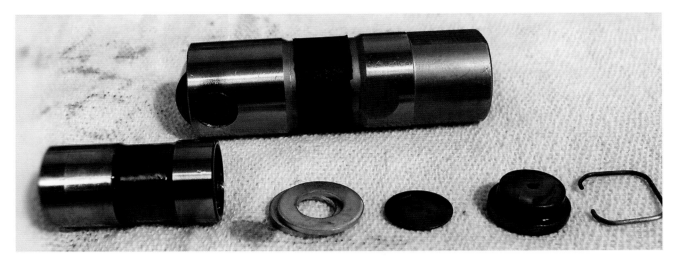

To make a solid lifter out of a hydraulic lifter, dismantle and set aside the small spring in the bottom. Then reassemble it like stock but use shims or washers under the pushrod cup to take up the slack. It is helpful to mark the lifter with engravings to keep it separate from the usable lifters.

Now that the lifters are solid, you need to mark them so you don't confuse them with your usable lifter set. I have used black paint and markers, but more recently, and to quickly identify the "solid" lifters, I began using a short length of 3/4-inch heat shrink as a band on the oiling recess. This heat shrink band is thin enough to allow the lifter to move in the lifter bore without binding it.

Install your new hybrid "solid" lifters into place on the cylinder no. 1 lifter lobes with the camshaft on its base circle for that cylinder. The base circle on cylinder no. 1 is when both the crankshaft gear alignment dimple and the camshaft timing gear dimple are both straight up at 12 o'clock. Next you can do one of three things to hold the lifters into place: Install the factory lifter tray that clamps the lifters tightly. Install no tray and pray the lifters do not rotate, or use an OEM-type tie-bar system from another engine. I unintentionally discovered that the plentiful small-block Ford steel lifter guides fit when ground slightly for block clearance and will hold the lifters properly oriented into place, keeping the lifter rollers perfectly in line to the camshaft lobes. If careful, you do not need anything to keep the lifters aligned because when they are rotated slowly with light pressure against them, they tend to align themselves.

Install the lightweight checking springs onto the first cylinder head that you intend to install onto this engine. The lightweight springs allow the use of light hand pressure to open the valve during the PtoV procedure. Remove the intake and exhaust valve springs. Reusing the retainer and valve keepers, install the checking spring. Next we will need to mock-assemble the cylinder head onto the block, using a head gasket of the thickness you intend to use in the assembled engine. Install the head gasket first, then the cylinder head, and at least two cylinder head bolts adjacent to the no. 1 cylinder to hold the cylinder head in place. The bolts do not

need to be torqued down, just fitted close enough that the cylinder head doesn't move and is firmly in place.

Most MLS gaskets are close to their compressed thickness when not compressed. If not, remember to subtract the difference between your uncompressed thickness and compressed thickness from your final PtoV readings as you will have slightly less clearance. Usually, the max difference here is about 0.005-inch, so if you are close enough that 0.005-inch makes a difference anyway, something drastic needs to be done. Install your adjustable length pushrod at its shortest length into the intake pushrod hole first, and then install the rocker arm system, tightening the rocker arm on the valve you are checking clearance on.

To check PtoV, we want the engine as close to running condition as possible. Use the cylinder head you are going to run and the exact same head gasket.

Next, install the dial indicator before you adjust the pushrod or rocker arm. If you have a magnetic base indicator, you will need to affix a metal plate onto the cylinder head for the magnet to adhere to, or you may procure a long 8-millimeter bolt and use that as your dial indicator base fixture. If you get the 8-millimeter bolt, fasten it into a nearby unused rocker arm bolt hole, so you don't have a ton of deflection in the dial indicator setup. Adjust the dial indicator so the finger of the dial indicator is preloaded onto the valve retainer. Remember that if you have 0.6xx-inch lift, you need to have the dial indicator preloaded at least that same amount, which is well past halfway of the travel length on a 1-inch dial indicator. Leave enough room to manipulate the valve retainer or rocker arm by hand when you depress the valve in the next few steps.

With the pushrod, dial indicator, and rocker arm in place, take the extra slack out of the pushrod by extending it in length until it rests against the rocker arm. You will want the valvetrain to be at 0 lash and 0 preload to have the highest accuracy with your PtoV measurements. Adjust the length of the pushrod while watching the dial indicator needle to make sure you are not adding lift. You may preload the valvetrain slightly but no more than 0.001-inch preload so as to not skew the results greatly. Once 0 lash/0 preload is achieved, you can zero out your dial indicator, but this is not necessary as PtoV issues have no reference from 0 lift. You are measuring from an "open valve" point of reference until piston physical interference, with both changing constantly at different points of lift and piston travel. You will have to zero the dial indicator each time PtoV is checked at each degree point.

A lightweight test spring is needed to manipulate the valvetrain for dial indicator measurements.

Using an adjustable-length pushrod, zero out the lash/slack in the valvetrain while the lifter resides on the base circle of the camshaft. For cylinder no. 1, this is when the camshaft gear mark would be at 12 o'clock. Install the dial indicator on the valve spring retainer as we are measuring valve clearance.

If no dial indicator is available, the other option is to use a set of feeler gauges. Using the same 15 BTDC to 15 ATDC range, use a stack of feeler gauges that measures 0.080 inch on the intake, and 0.100 inch for the exhaust, to verify the minimum clearance is obtained. The measurement location is between the valve stem tip and the rocker arm friction point. Turn the engine to the desired location, push the valve down against the piston, and insert the feeler gauge stack. Do not rotate the engine over with the feeler gauges under the rocker arm.

Alternatively, if no adjustable length pushrod is available, you may tighten down the rocker arm bolt just enough until no lash is in the valvetrain. This makes for sloppy side-to-side movement, but it's adequate for PtoV measurements if done carefully. Remember to always zero the pushrod while the lifters are on the base circle of the camshaft for a no-lift point of reference. Similarly, if an adjustable rocker arm is used, no adjustable pushrod is required. Just adjust the lash out of the valvetrain using the rocker adjustment bolt.

The next procedure is easiest to document with the degree wheel in place. We want to start turning the crankshaft over until

we are at the overlap point of the camshaft. The rough middle of overlap is 360 crankshaft degrees from the base circle location. You want to sneak up on the PtoV clearance when using solid lifters. If you do have a mechanical interference problem and you turn the engine over, you may bend the valves that hit. Turn the crankshaft over until the dial indicator starts to see lift and start measuring right away once valve lift is noticed. Note the measurement of how much the valve travels before it touches the piston by lightly manipulating the rocker arm away from the pushrod to simulate more lift. The measurement on the dial indicator is your piston-to-valve clearance.

The general point where intake valve PtoV clearance needs to be measured is approximately 10 degrees before and after TDC, although I would recommend measurement increments of 5 degrees from 15 degrees before to 15 degrees after. You want to find the tightest clearance. Repeat this exact setup and procedure for the exhaust valve to measure exhaust PtoV clearance. Minimum recommended clearances are 0.080-inch intake and 0.100-inch exhaust. You can get away with a little less at times by exercising caution and using a good valvetrain.

Rotate the engine over until you are 15 degrees BTDC and take your measurements on the intake valve in 5-degree increments through 15 degrees ATDC. Determine the clearance at each step by pushing the valve until it hits the piston, zeroing out the dial indicator when it hits the valve.

Next, let pressure off the valve and where the dial indicator stops is your PtoV clearance for that degree point. This intake clearance was at 0.121 inch. The recommended minimum is 0.080 inch for a hydraulic camshaft. Note that peak interference occurs when the valve is at 0.275 inch of lift, not even close to peak lift of 0.630 inch.

With modern aggressive cam profiles you will find that the tightest point of intake valve PtoV stays consistent for a few degrees of crankshaft rotation ATDC as the piston out-accelerates the valve down the cylinder. Camshaft engineers design the profile of the camshaft lobe so that the valve will open at the same speed that the piston is traveling down the cylinder after TDC for maximum valve opening speed without kissing the piston. Typically, the quicker the intake valve reaches peak lift, the more air and fuel charge will fill the cylinder, restricted to the physical limitations of the internal engine clearances. The speed of the valve chasing the piston, while maintaining adequate PtoV, is the limit of the valve's opening speed.

Compared to OEM engines, most built LS-series engines will have a ton of valve clearance. The main variable factors are the amount of cylinder head milling, piston valve pocket depth and diameter, valve size, and camshaft size.

Document your minimum PtoV measurements on your engine build sheet.

If you have less-than-desired clearance, there are a few Band-Aids that may help. If the intake valve has too little clearance while the exhaust has a ton, you can retard the camshaft timing a few degrees to help out. Likewise, if the intake valve has plenty but the exhaust is lacking, you can advance the camshaft timing a few degrees. Recheck PtoV after any camshaft timing changes to verify. Keep in mind that this adjustment will affect the power band and may affect total power output if non-desirable valve events come out of this Band-Aid change. If both intake and exhaust are tight, you can widen the lobe separation angle and retard the camshaft at the same time, but this can only be accomplished with a new camshaft grind. If such significant changes are required, consult with your camshaft manufacturer for recommendations.

Using the same procedures but on the exhaust valve, check PtoV at three points BTDC. Using hand pressure, rest the exhaust valve against the piston and zero the dial indicator.

Let the valvetrain slowly relax while taking note of the dial indicator reading. This clearance is 0.140 inch, which is well above the minimum recommended clearance of 0.100 inch for the exhaust valve. This would indicate that another 0.040 inch of head milling is optional.

One thing that may help on any factory short-block is the addition of valve reliefs. ISKY and a few other companies offer piston-notching tools and a drive mandrel that fits through your valve guide. Because the factory pistons do not have valve pockets, this allows you to carefully cut your own. Some people will perform this fly cut using their new heads, but to properly fly cut with the ISKY tools, the valve seat needs to be removed to put the cut in the right location on the piston. Without removing the valve seat, the reliefs are shifted into a position that doesn't help PtoV clearance. The work is tedious the first time, as you need to set up the depth of cut, but once set up the pistons can be modified in two to three hours. Keep a shop vacuum nearby and block off the lifter area with shop rags as this can get messy. Ideally it is best to fly cut with the engine removed from the vehicle.

The other easy thing to change is to run a thicker head gasket. If you are running a 0.040-inch-thick gasket, swapping to a 0.051-inch gasket nets about 0.010 inch to both intake and exhaust clearance. One other last resort is to have the valve job in the heads sunk a few thousandths of an inch deeper than standard. This is typically limited to 0.015 inch of extra depth. So if your PtoV clearance is 0.030 inch, this won't be the magical fix. This procedure also tends to affect cylinder head flow after a certain point. Only skilled cylinder head specialists should ever attempt this. The good news is that most aftermarket casting heads typically have the same or better PtoV clearances as stock heads. Only move on from this point once you know your clearance measurements are accurate and sufficient. Always double-check your measurements.

Remove your solid lifters and lifter guide, and set them aside with the rest of your camshaft degreeing and measuring tools. Don't make the mistake of using your "solid" lifters and assemble the engine around them.

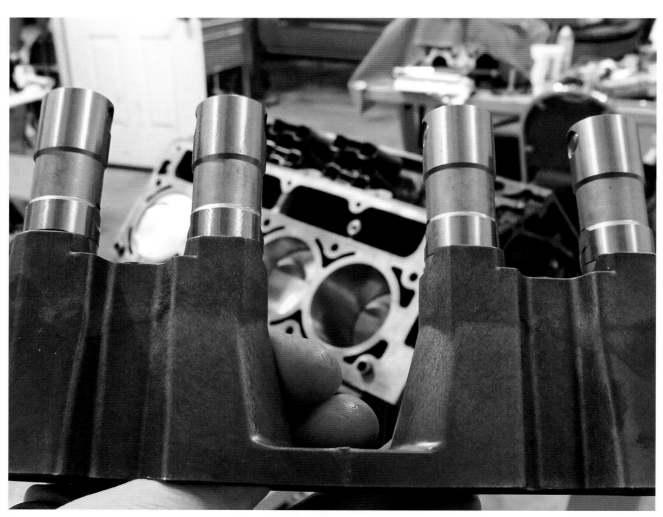

Now that the camshaft is degreed and the PtoV clearance is verified, we can once again throw some parts at the engine. Install the lifters into the lifter trays with light oiling. The oiling hole location doesn't matter, but I install them all to the rear of the engine.

INSTALLING LIFTERS AND LIFTER TRAYS

After degreeing in the camshaft, you'll want to install the new lifter set. Prepare the lifters by cleaning the outside of the lifters with some solvent. At the same time, make sure the roller turns freely and there are no blemishes, cracks, or nicks in the lifter body itself. If everything looks good, air dry the lifters and soak them in clean engine oil for at least 30 minutes before assembly. After soaking the lifters, you will want to install them into the factory lifter trays. There are four lifters per tray and four trays. If you are using a Gen IV block, you will need lifter trays that are notched to clear the displacement-on-demand (DOD) lifter location bosses. These really have no function other than to mark the location of the DOD lifters when used. If using tie-bar lifters, you will not need the factory plastic lifter trays to keep the lifters aligned. Many people ask which way the lifter oiling feed holes are intended to be aligned. The location does not matter, but I like to align them all the same to keep uniformity. If the oiling hole is on the side of the lifter body, I typically point these all toward the rear of the engine. There is no right or wrong way, as the lifters are encircled within constant oil pressure from the oil galleries.

Using clean engine oil or assembly lube coat the lifters and make sure the rollers spin freely. These particular lifters are pre-lubed so no soaking is required, although soaking the lifters in engine oil never hurts prior to assembly.

The extra assembly lube on the lifters will help lubricate the lifter bodies upon engine startup, until engine oil pressure arrives. The assembly lube also assists lifter installation.

Once the lifters are all installed in the plastic lifter trays, you can add further engine assembly lube to the roller portion and side of the lifter body and then install them into the block as an assembly, four lifters and one tray all at one time. You will have to exert slight side-to-side movement as you align all four lifters with their corresponding lifter holes in the block. Once installed, torque the four-lifter tray retaining bolts to 106 lb-in. This is a specific and unique shouldered bolt as it lets the lifter tray move around slightly so as not to bind any lifters in place as they are opening and closing.

INSTALLING CYLINDER HEADS

Next up for installation are the cylinder heads themselves, but before installing the heads there are a few small items to take care of. First is installing the four-cylinder head alignment dowels, if they have not been taken care of already. This is a straightforward installation using a small hammer to carefully tap them into place. The alignment dowels have a matching recess in the block on the lower four corners of the deck surface. These protrude through the cylinder head gasket to hold it in alignment. This also will hold the cylinder head itself in place. The gap in the dowel can be aligned in any

After installing the lifter trays with lifters, install the tray bolt, which allows the tray to move slightly to prevent binding as the bolt is torqued down.

Loctite is not needed on these bolts, although it never hurts either. Torque the lifter tray bolts to 106 lb-in (9 lb-ft).

Install the four cylinder head locating dowels by tapping them into place with a small hammer. If reusing your old dowels, they might need to be expanded slightly to fit tightly.

If using an ARP head stud kit, now is the time to install the stud portion of the kit. Simply thread these in by hand until they bottom out. It's much easier to install head studs in new blocks because they have never had stock bolts with epoxied threads. Use of an ARP M11 thread chaser is recommended on used blocks.

direction. Like the lifters, I tend to point them all toward the center or all upward. It does not matter much.

Depending on if you are using cylinder head studs or bolts, you will want to install the threaded portion of the head stud kit into the block. Make sure the threads are clean by either using an 11-millimeter thread chaser or a modified old head bolt first. Normally, new threaded studs on a clean block will thread in by hand pressure. Some used blocks, which cannot be 100 percent cleaned such as a new block would be, may need a little extra help by employing an Allen-head bit to thread the stud into place. Nothing other than a light film of oil is required on the threads that protrude into the block, as these are blind holes that do not extend into the cooling system. These studs do not get torqued down themselves. All that is required is for them to be installed fully by light hand pressure, if the threads are clean enough to tighten by hand.

I cannot stress enough that these blind head bolt holes need to be cleaned thoroughly. Old head gaskets, dirt, machining grit, and any other foreign debris will negatively affect head bolt torque readings and often damage the threads at the same time. If absolutely any coolant or fluids are left in these blind head bolt holes when you tighten the head bolts, these fluids have nowhere to go and will usually crack the block. This is a more common problem when doing a cylinder head swap in your car, as engine coolant will trickle into the lower set of head bolt holes when the old heads are removed, but this is of importance enough to be worth mentioning for any buildup and will be mentioned throughout the text.

On all LSX-based production blocks, there are 10 main M11 cylinder head bolts per engine bank and

5 smaller M8 bolts along the intake flange near the valley plate. With aftermarket type six-bolt blocks, the same OEM fastener location is provided so that any production-based cylinder head will physically fit if needed, but these have an additional eight-bolt hole locations per bank in between the main M11 bolts for added strength and further help with head gasket sealing on huge power-adder setups. With the GMPP LS-series block, the outer bolts are small M8 head bolts much like the production M8 bolts.

Now that the head bolt holes are cleaned, and the head studs are installed if used, the cylinder head gaskets can now be installed. Some manufacturers offer a separate left and right bank gasket, such as the graphite OEM head gaskets, and some have the same exact part number on the left and right bank, with one installed "upside-down" from the opposite side. Most head gaskets are marked "Front" and bank-specific gaskets will be marked "Top" to alleviate installation confusion. If they are not marked, remember that the larger coolant holes of the head gaskets will align to the rear of the block, near cylinders no. 7 and no. 8 on each bank.

Coolant is intended to travel from the water pump, through the length of the block first, and then up into the cylinder head past these two coolant passages in the rear of the cylinder head gasket, where it then changes direction and travels back through the cylinder head and back to the water pump and into the radiator. If the gaskets are installed wrong, these coolant holes will be restricted, possibly leading to engine overheating and hot spots throughout the cooling system.

Put the head gasket into place, making sure to align the rear coolant holes in the block with the coolant passages in the head gasket. It will physically fit either direction. When using Cometic MLS head gaskets, it is sometimes necessary to coat the gasket with copper gasket sealer to prevent coolant seepage.

With the proper head gasket in place, gently set the cylinder head into place over the studs or head dowels.

If using non-OEM multi-layer steel (MLS) head gaskets, then carefully consider the block deck surface roughness and cylinder head decking finish. Cometic Gasket gaskets require a surface finish of 50 roughness average (RA) or better to seal correctly on both the block and cylinder head surfaces. This would need to be checked at your machine shop, but a quality resurfacing job on both the block and cylinder heads is usually well within adequate range. These MLS stainless steel gaskets have a 0.001-inch-thick Viton coating that will compensate for minor irregularities in the deck surfaces. If you have coolant sealing issues using Cometic MLS gaskets, you may need a light coating of sealant on the head gasket to fill in these gaps. Some builders use silicone around the water ports, while others have good success using two light coats of copper gasket spray. Aftermarket MLS head gaskets can be had in almost any thickness (usually in 0.005-inch increments), while OEM head gaskets are only available in stock compressed thicknesses of 0.055 inch or 0.060 inch.

Make sure the deck surface is clean of any oils or solvents. You may now place the head gasket of your choice onto the block. If studs are used, you need to evenly work the head gasket past the studs until it is in place on the block aligning with the coolant ports discussed above. Once the head gaskets are in place, you may install the cylinder heads themselves. They are not normally side-specific. Certain later model LS7 heads are side-specific due to the rear coolant bleed holes not being fully machined from the factory. This practice may appear in future Gen IV engine designations. If the coolant bleed holes are not drilled, make sure to install the heads with the drilled and machined bleed holes to the front.

Now that the heads are installed, we can bolt them into place. If using head studs, now is the time to install the hardened washers and hardened 7/16-inch matching nuts to the head studs. Before installing these, apply a mild coating of ARP moly lube to the threads and washer faces of the nut and washer, then install the nut using a 1/2-inch

When using ARP hardware, it is necessary to use the proper ARP moly lube to achieve accurate torque values. Apply a mild coating on the hardened ARP washer. This one was a little too generous but will work nonetheless.

12-point socket and extension for 12-point ARP nuts, and a 5/8-inch six-point socket and extension for six-point ARP nuts. If using ARP head bolts, the same technique applies. Use a 1/2-inch six-point socket instead. Apply ARP moly lube to the bolt threads and to the hardened washer face of the ARP washer. Remember to note that the chamfer on the inner diameter radius of the ARP washer needs to face toward the hex portion of the ARP nut and bolt as there is a radius under the nut or bolt head to increase strength, while the flat portion of the washer should be against the cylinder head itself.

This chamfer clears the under-head bolt radius much like the rod bearing chamfer clears the radius on the rod journal as discussed in the short-block chapter.

After installing the main head fastener hardware, you can install the remaining five small M8 bolts or nuts on each bank.

If using stock OEM type fasteners for the head bolts, remember that every time these are torqued, they will need to be replaced. The OEM M11 head bolts are throwaway bolts and are never intended to be retorqued. These are straightforward to install; just install and torque them out of the box. You may use ARP moly lube on the washer face to help with the strenuous GM torque procedure up next, but no lubricants are necessarily required on the threads, as they have an epoxy thread compound that locks the head bolt in place after it is torqued completely.

Install all hardened washers first with the chamfered inner edge facing upward. It does not matter much for head studs, but for head bolts, not locating the chamfered edge correctly will cause interference and improper torque values.

Much like the ARP hardened washer, apply a moly lube coating to the washer face of the ARP threaded nut and a smear inside on the threads themselves.

TORQUEING THE CYLINDER HEAD FASTENERS

If using the ARP head stud kit, torque the fasteners in three steps up to the total torque value. ARP recommends torqueing the ARP nuts to a total torque of 80 lb-ft on the M11 studs. Follow the OEM torque sequence bolt locations in order, using a first step of 30 to 35 lb-ft, second step of 50 to 55 lb-ft, and a final step of 80 lb-ft. After doing the final torque of the M11 studs, it's beneficial to retorque them all to 80 lb-ft. The first few bolts will have less clamping force than the last few bolts once the entire head is torqued, because you compress the head gasket in steps. With the small M8 head

studs on the top row, torque these in a final and last step to 22 lb-ft. No additional torque or angle procedure is required for these M8 head bolts. Follow the same procedure on the opposing bank's cylinder head.

When using ARP head bolts, the same procedure is followed with the only difference being the final torque values. Torque the M11 bolts in sequence and torque step values in the following manner: 30 to 35 lb-ft, 50 to 55 lb-ft, and a final pass of 70 lb-ft. As with the head studs, a final retorque of the M11 bolts is recommended. Retorque the bolts to 70 lb-ft. You will find at least the first half of the bolts will be

I find it best to install each ARP nut with a properly sized socket and extension. Thread each nut into place until it bottoms out on the washer by hand.

looser than the last few bolts torqued in the sequence. Once all M11 bolts are final-torqued, you can move to the upper row of M8 head bolts and torque these to 22 lb-ft. Follow this same procedure on the opposite side.

If using OEM fasteners, the torque process is dissimilar to what many would consider "normal" torque procedures. These fasteners use a base torque setting as a baseline and then a stretch value achieved by angle-torquing the bolts a set amount of rotation degrees. This procedure permanently stretches the bolts and provides head gasket clamping via the tensile strength of the stretched bolts. Once the angle torque process is complete, each head bolt is stretched approximately 0.040 inch. There is no shortcut method to torqueing these bolts through stepped tightening techniques, as when these bolts are stretching, the same high-torque value is applied consistently until the desired angle is reached. Prepare for a workout when torqueing these OEM fasteners. It's quite a strenuous process, and more so when the engine is installed in a vehicle and a cylinder head swap is being attempted. Allow extra time for recuperation and rehydration after this procedure. You will quickly learn to dislike the torque-to-yield head bolts.

The torque procedure starts in the center and works outward. Torque the M11 ARP hardware in three steps of 30 lb-ft, 50 lb-ft, and 70 lb-ft for ARP head bolts or 80 lb-ft for ARP head studs. Torque the top row of M8 bolts torque to 22 lb-ft as the final step.

Here is the LS-series recommended torque pattern to be used with GM OEM hardware: Similar sequence pattern can be used with ARP hardware.
First step: no. 1 to 10 = 22 lb-ft
Second step: no. 1 to 10 = 90 degrees
Third step: no. 1 to 8 first design long bolts = 90 degrees
Third step: no. 9 to 10 first design medium bolts = 50 degrees
Or third step: no. 1 to 10 second design medium bolts = 70 degrees; no. 11 to 15 = 22 lb-ft

First torque the 10 M11 OEM bolts to 22 lb-ft following the standard torque sequence procedure, and then retorque them again to 22 lb-ft to make sure all bolts start off at the same base tightness before angle torqueing. I recommend marking the OEM head bolts with a black marker after the initial torque sequence is finished with a line or dot in the 12 o'clock position. This will help you keep track of which bolts have been angle torqued. Some use this mark to reference the angle torque itself if not using a torque angle gauge. Next you will need a simple angle torque gauge. You can pick up an inexpensive gauge at your (maybe) friendly neighborhood parts store or Internet vendor for about $20 or from one of the large tool vendors like Snap-On for a little more. Simple angle gauges are all the same.

In the second torque step you need to angle torque all the M11 bolts, turning each bolt 90 degrees in sequence. If using the first-design long M11 bolts, rotate these an additional 90 degrees as the third step. The last two medium-length bolts get 50 degrees additional rotation instead of the 90-degree step. These medium-length bolts are in positions 9 and 10 of the torque sequence. If you have the second-design head bolts, which are all "medium" length, the third torque step is 70 degrees instead of 90. Finally, the M8 bolts torque to the same old 22 lb-ft. Repeat this procedure for the opposite bank.

I know it may sound confusing for those used to conventional torque settings. To attempt simplification for reference:

1. Torque M11 bolts to 22 lb-ft.
2. Torque M11 bolts 90 degrees.
3. Torque M11 bolts an additional 50 or 90 degrees depending on bolt length, or torque all second-design head bolts to 70 degrees.
4. Torque all M8 head bolts at intake runner flange to 22 lb-ft.

INSTALLING PUSHRODS AND ROCKER ARMS

After installing and torqueing the cylinder heads and bolts, the remainder of the valvetrain components may be installed. But first a few measurements are necessary. Before installing the pushrods, we need to know what length they should be. This affects lifter preload, and preload amount varies with manufacturer and type of lifter. Typically you would want to duplicate OEM measurement amounts, but a few high-rpm type lifters require close to zero preload. To measure required pushrod length you need an adjustable pushrod length checker and a dial indicator for accurate measurement of preload.

Normally, just checking one cylinder is adequate. The easiest to check is cylinder no. 1 or cylinder no. 6, as the timing chain marks correspond to these two cylinders. With the camshaft gear's dot at 12 o'clock and the crankshaft dot at 12 o'clock, you can check pushrod length or preload on cylinder no. 1. With the dots aligned (camshaft at 6 o'clock and crankshaft at 12 o'clock) you can check pushrod length or preload on cylinder no. 6. This position is the firing position, where both camshaft lobes are on the base circle of the camshaft, so both intake and exhaust pushrod length and preload may be checked at the same location.

Install the pushrod length checker into the intake or exhaust pushrod hole and install the rocker arm assembly for that cylinder or the entire assembly if using stock rockers. Tighten the rocker arm down, and just start a few threads on the rocker arms that you are not checking in order to keep the assembly aligned. You'll need to tighten or torque the rocker arm on which you are checking pushrod length. Extend the pushrod length checker to meet the rocker arm. This is a little hard to do as the whole pushrod will want to turn. Depending on the cylinder head design, if you can't extend the pushrod with the rocker installed, you may need to remove the rocker arm, extend the pushrod a little, and then reinstall the rocker until the length is determined through trial and error.

Usually pushrod length checkers are marked in 0.050-inch increments and have a base measurement marked on them. For example, with a checker with a base length of 6.800 inch, if you have eleven 0.050-inch stepped line indicators showing, the pushrod length would be 7.350 inches at zero lash. But because most hydraulic lifters need preload, you add your desired preload to that number to determine actual required pushrod length. If you need 0.075 inch of preload, you add 0.075 inch to your measured 7.350 inches and come up with 7.425-inch pushrod length. Repeat this process for the opposing valve and opposite engine bank. Normally it should be extremely close if both decks are the same height and both cylinder heads are decked the same.

Once proper pushrod length is determined, you can install the set of pushrods and then the rocker arm system. Install the pushrods with a dab of assembly lube at either end for start-up lubrication. Inspect the intake rocker bolt holes to see if they go through to the intake port itself. If so, you will want to use sealant on the intake rocker bolts so as not to pull oil through the bolt threads and into the intake runner. Use either oil-resistant gray or black RTV silicone for best results. Depending on rocker arm type, you can either install the entire system at once or rocker arms paired by individual cylinders.

Continued on page 152

PUSHROD LENGTH

There has always been a lot of discussion on proper pushrod length and many viewpoints on how to calculate pushrod length without actually measuring what your engine needs. While calculating pushrod length by measuring the base circle of the camshaft and comparing it to stock will get you in the ballpark with a 100 percent stock engine, when you start changing other things, there is no absolute point of reference anymore. Changing camshafts obviously changes pushrod length as when you gain lift, usually the base circle of the camshaft is nominally smaller. In addition to camshaft lobe design, other simple things like lifter length, valve length, valve-job, cylinder head/block milling, head gasket thickness, and rocker arm design all have a say-so. When you start changing more than one of these parts, then pushrod length calculations without physical measurements are completely skewed.

The preferred method for checking for proper pushrod length is to use a pushrod-length checker on both a closed intake and exhaust valve. The reason for checking both is because many cylinder heads have different height intake and exhaust valves. That difference may not be absorbed by the lifter plunger itself.

First we need the no. 1 lifters on the base circle of the camshaft. If you still have the timing cover off, you can see that this is with both timing marks at the 12 o'clock position. Install the pushrod length checker at a base setting first, and then install the intake rocker arm. Tighten the bolt finger tight until it bottoms out. Try to move the rocker. If there is a ton of movement, extend the rocker in 0.050-inch increments until the lash is removed. You will likely have to remove the rocker and reinstall it for each increment. The goal here is to have 0 movement on the rocker arm without preloading the lifter or the intake valve itself.

Once the 0 lash setting is found, remove the pushrod and measure it. The Competition Cams pushrod length checker I use has hash marks at each 0.050-inch increment with a base setting of 6.800 inches. On this particular engine, we have 10 turns past the base setting of 6.800 inches, 10x0.050 inches = 0.500 inch, which we add to 6.800 inches for a 0 lash setting of 7.300 inches. Now this isn't the pushrod we need. Depending on how much preload we want in the lifter, we add the desired preload to this measurement. For these OEM lifters, I prefer to use 0.050 inch to 0.075 inch of preload. Add this to our measurement and we're at a 7.350-inch pushrod length. Verify the exhaust pushrod length, as it may require a different dimension. The engine I was working with here was within 0.010 inch of the intake length, which is close enough to use a pushrod of the same length.

I realize that some people may not have a pushrod length checker or may be in a bind trying to quickly do a camshaft swap in their driveway or taking up a buddy's garage spot. I didn't forget about these guys; there are a few measurement alternatives to using an adjustable pushrod.

The first method is using a dial indicator on the pushrod side of the rocker arm and using some existing known-length pushrods. If you have some pushrods that are close to what you think you will need, we can use these. Typically LS pushrod length on Gen III OEM heads is 7.350 to 7.425 inches, with 7.400 inches being close to stock length. Since 7.400 inches is the most common, I will use this length for the sample measurement.

Ideally, to check required pushrod length or lifter preload, an adjustable pushrod should be used first. The pushrod length checker is extended until it meets the locating cup in the rocker arm. With no lash and no valve opening, this is zero lash and zero preload.

If using "replacement" rocker arms where the intake and exhaust are paired together, this measurement must be taken individually per specific intake or exhaust and the mating rockers must be split up for a few moments.

Most pushrod length checkers will use a base measurement with each rotation of extension adding a certain amount of length. In this sample case, the base measurement is 6.800 inch. Add 10 rotations of 0.050 inch each = 7.300 inches to achieve zero lash/preload. Add your desired preload to this number, and that is your required pushrod length.

When using ported heads, apply a dab of RTV sealer to the intake rocker bolt threads and install.

An alternative method to measuring pushrod required length is to use a dial indicator on the pushrod side of the rocker arm while using a known-length pushrod. Preferably, use one that is close to the right length. Tighten the rocker bolt by hand until it is against the pushrod at zero lash and zero the dial indicator.

Then, with the lifter on the base-circle, tighten the rocker bolt until it bottoms out. When the dial indicator stops moving, that is your actual lifter preload. If the dial indicator does not move much at all, then you probably need a longer pushrod. Alternatively, if the reading is too much preload, then a shorter pushrod is required.

Install your known-length pushrod into the no. 1 intake position with the intake lobe on the base circle as in the previous measurement, and set the rocker arm in place. Tighten the rocker arm until you have no slack in the rocker arm and it is against both the lifter and pushrod with extremely light pressure. This is zero lash. Now you will need a dial indicator. Set up the dial indicator to measure the pushrod side of the rocker arm and zero out the dial indicator. Now tighten the rocker arm until its bolt is bottomed out and the rocker is tight. Note the measurement of the dial indicator. Remember the dial indicator is extending, so the number is decreasing, unless you have a dual-pattern dial indicator gauge.

Your measurement is your preload with the known-length pushrod. If the dial indicator did not move much, then you need a longer pushrod. For the sample measurement, my readings were 0.075 inch of preload using a 7.400-inch pushrod. If less preload is desired, I could go with a 7.375-inch length and achieve 0.050 inch of pushrod preload; but since this is close to how I normally set up the valvetrain, we'll leave it alone for now.

The next thing to check is the rocker wipe or contact pattern if using full roller rockers. Ideally this is checked with solid lifters, but may be checked with the lifters that are already installed in the engine. The goal here is to have the rocker pivot close to the center of the valve during operation. Color in the tip of both the intake and exhaust valves on the cylinder you are checking. I use a black marker and have found out that the dry erase markers work well also. After the marker ink dries, carefully install the rocker arms and torque them to specifications

(24 lb-ft). Now quickly rotate the engine over two full rotations and stop where you started.

Next, remove the rocker arms and inspect the contact pattern. We're looking for a centered wipe pattern. In the photos nearby, my first measurement was close to the intake manifold side of the rocker tip, which means the rocker geometry is off a little. This can be due to the valve location in relation to the rocker location. In this case, a shim is necessary under the rocker arms. By adding a 0.045-inch shim under each rocker arm pedestal, we are able to correct the geometry issue easily and obtain a correct wipe pattern. Changing pushrod length does not affect rocker wipe unless the pushrod length was way off in either direction to where the lifter was bottomed out or there was valvetrain lash. The lifter compensates for pushrod length to an extent. If you have a 7.350-inch pushrod or a 7.400-inch pushrod and have preload with either, rocker wipe will be identical with both.

Note that adding a 0.045-inch shim changes the ideal pushrod length. We now have to go back and check pushrod length to verify it, and probably will need to go with a slightly longer pushrod to compensate for the shim thickness. Engine building is a cause-and-effect game throughout.

When using factory rockers, this is one area of concern, as there isn't a whole lot that can be done to achieve the correct rocker wipe pattern. Past 0.550-inch lift, the rocker arm is dragged across the valve stem tip until peak lift is obtained, resulting in a wide contact patch. When using the harder OEM-type metal valve guides with stock rockers, this wipe issue isn't of much concern, but bronze valve guides will wear out. It's not a question of *if* they will wear, but *when*.

The rocker wipe pattern is something many LS people tend to overlook, but it is something that needs to be checked on non-OEM cylinder heads when using aftermarket rocker arms. The first thing to do is to take a black marker and color in the tip of the valve stem.

Then carefully install the rocker arms and rotate the engine over a few times to generate a thorough rocker wipe pattern. This type of non-adjustable fulcrum mount rocker arm can only adjust its rocker wipe geometry by using shims to change rocker arm location.

This rocker wipe pattern is concentrated close to the intake flange side of the valve, which is a normal effect with ported and non-OEM heads. The addition of shims under the rocker pedestals should correct this anomaly. These shims are 0.045-inch thick, which also will affect pushrod length adversely. Recheck your pushrod length if rocker shims are required.

After the addition of rocker shims, the rocker wipe pattern is concentrated in the center of the valve tip. The wipe pattern width can also be measured. The narrower the better, but other than fine shimming the rockers or using fully adjustable stud-mounted rockers, not much adjustment of this can be achieved.

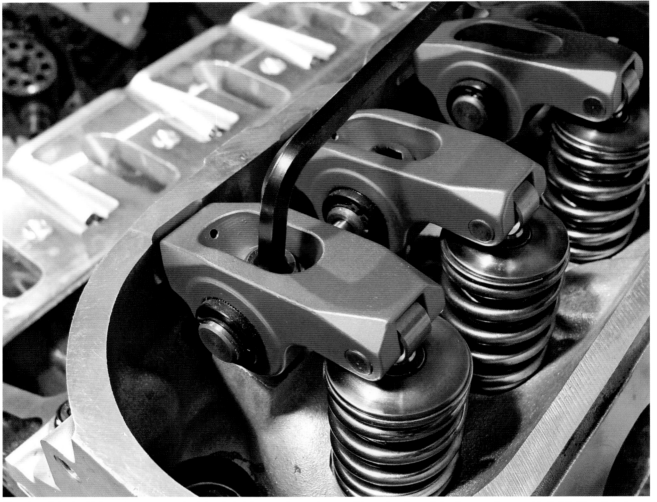

When the cylinder head is torqued, we can move back to valvetrain items such as rocker arms. When using aftermarket rocker arms, these must be installed by cylinder in pairs while both cylinder lifters are resting on the base circle of the camshaft. If you know your pushrod length, you may install the rockers now.

Just like any other internal engine part, there is a set procedure for tightening rocker arms. You want to tighten and torque individual rocker arms while the lifter is on the base circle of the camshaft. With stock rockers, you can tighten the intake or exhaust individually, while with aftermarket setups, you usually need to do both as a pair per cylinder.

The procedure for tightening stock rockers is to first put the engine at TDC for cylinder no. 1 and note where the upper cam gear alignment dot is. Each time cylinder no. 1 is at TDC, half the lifters are on the base circle or close enough to being closed to not matter. Make sure the rocker you are torqueing does not open the valve a significant amount as you are tightening it. Factory specification for the rocker bolts is 22 lb-ft, but with an aftermarket camshaft, springs, pushrods, and a higher rpm limit, it is a good idea to bump it up a tad to keep rocker arm bolts from backing out; 24 to 26 lb-ft is adequate.

FACTORY ROCKER ARM TORQUEING PROCEDURE

If the camshaft gear dot is at 12 o'clock or TDC firing position for no. 1, tighten and torque the intake rockers on cylinder's no. 1, 3, 4, and 5, and exhaust rockers of cylinder's no. 1, 2, 7, and 8. Turn the crankshaft one full revolution until TDC is achieved again. The camshaft alignment dot should be at 6 o'clock. Now the other half of the rocker arms may be tightened down and torqued to specification.

AFTERMARKET ROCKER ARMS

Most bolt-on aftermarket rocker arms need to be tightened by pairs instead of individually. For non-adjustable rocker arms, the easiest way is to install and adjust each rocker arm pair when that particular cylinder is at TDC firing position, where both valves are closed. If the timing cover is still off, you can simply start at cylinder no. 1 and go through each cylinder in sequence according to firing order (see the firing order below). Some rocker arms need to be shimmed, or different length pushrods need to be used. Go back a few paragraphs to the pushrod measuring section or reference the preload section if needed. Shimming the rocker arms is almost like adding a longer pushrod, although doing so changes the rocker arm wipe pattern on the valve tip, whereas shorter or longer pushrods do not. See the sidebar

in this chapter for more information on adjusting the rocker arm and valve top contact pattern.

To start, rotate the engine until both timing chain dots are in the 12 o'clock position, both pointing upward. This is the TDC firing position for cylinder no. 1 yet again. Install the rocker setup, using RTV silicone on the intake rocker bolt if required due to the intake bolt protruding into

If reusing OEM rockers, the entire rocker rail and rockers may be installed at one time, although torqueing on each rocker bolt must be done when that valve is closed. Refer to the chapter text for location procedures.

Although the bolts on "open" rockers may not be torqued, you may start the threads and snug up these bolts by hand.

the intake runner. Evenly tighten the intake and exhaust rocker bolts until they fit closely and then torque both bolts to 26 lb-ft. Now rotate the engine one-quarter turn until the crankshaft gear dot is at the 3 o'clock position. The camshaft gear dot will be at the 1:30 o'clock position. At this point you can install the no. 8 cylinder's rocker arms. Follow the guidelines below to finish out the rocker arm installation

using the same procedures but with the engine rotated to the correct location for each cylinder. Each increment is one-quarter crankshaft turn, or 90 degrees from the last. It's an easy procedure to remember if you have the firing order on hand and a knowledge of valve events.

You may also use this guide to initially set the valve lash on a solid roller camshaft or for checking preload on

Following the recommended torque sequence and location, these OEM rocker bolts may be torqued to specifications.

OEM calls for 22 lb-ft, although I have found 26 lb-ft to be better when aftermarket components are used. Use care and do not tighten or torque a rocker arm when the valve is open. This results in skewed torque wrench readings.

individual cylinders. Use the numbers as if the camshaft and crankshaft gears are the hour hand on a clock. The crankshaft will make two revolutions per one camshaft revolution:

LS-Series firing order: 1, 8, 7, 2, 6, 5, 4, 3

- **DC cylinder no. 1**
  camshaft dot at 12:00 o'clock/crankshaft dot at 12:00

- **DC cylinder no. 8**
  camshaft dot at 1:30 o'clock/crankshaft dot at 3:00

- **DC cylinder no. 7**
  camshaft dot at 3:00 o'clock/crankshaft dot at 6:00

- **DC cylinder no. 2**
  camshaft dot at 4:30 o'clock/crankshaft dot at 9:00

- **DC cylinder no. 6**
  camshaft dot at 6:00 o'clock/crankshaft dot at 12:00

- **DC cylinder no. 5**
  camshaft dot at 7:30 o'clock/crankshaft dot at 3:00

- **DC cylinder no. 4**
  camshaft dot at 9:00 o'clock/crankshaft dot at 6:00

- **DC cylinder no. 3**
  camshaft dot at 10:30 o'clock/crankshaft dot at 9:00

When all 16 rocker arms are in place, it's good procedure to recheck the torque on all the bolts before putting anything else together. Anytime the rocker arm system is installed on an engine and there are new valvetrain parts, you need to pre-lube the top-end components of the cylinder heads with conventional engine oil, lubricating the rocker arm pivots, valve springs, valve seal area, and valve tip. It can take a few minutes of running for a steady oil supply to reach the valvetrain on a fresh engine. You want to make sure it survives those few moments when it's not getting oil. There isn't much more to do under the valve covers, so you can install the valve covers and torque them to specifications or move on to finishing out the bottom-end external components. This is also a good stopping point to take a break or set the engine aside for more assembly at another time if needed. The next chapter will cover installing the remainder of the oiling system and installing external covers.

After installing the rocker arm assemblies, it is a great idea to coat the valvetrain with the engine oil for startup purposes. It sometimes takes a few minutes of run time for oil to reach the valvetrain components.

Chapter 8
Finishing the Long-block

PREPARATION FOR INSTALLATION AND STARTUP

Reading over this text, you would think the engine assembly time takes weeks. Realistically, the assembly time from machined block to long-block is normally six to eight hours for a professional engine builder. If everything is prepared and ready, I've seen short-blocks assembled in less than two hours, including ring filing and bearing selections. If you are studying and following each step in the book while building your first engine, count on it taking longer.

Once the long-block is finished, all that's left is the oiling system, external engine covers, and a few engine sensors. I have documented the factory component install. There are aftermarket oil pans and timing covers available, but the installation does not deviate from stock.

First thing to install is the oil pump itself. The oil pump is mounted to the front of the block and driven by the lower timing chain gear. Pre-lube the oil pump with engine oil or thin assembly lube and line up the oil pump gears over the timing chain gear. If using a double roller chain, it will come with 3-millimeter-thick oil pump spacers for additional clearance to clear the extra width of the timing chain. Without the spacers the chain would grind into the pump, causing a few headaches. The spacer with two holes is for the passenger side (right side) of the engine. The spacer with three holes is for the oil pump outlet side of the engine, which is on the driver's side (left side). If using the oil pump spacer, a very, very thin smear of RTV silicone is needed on the oil transfer hole in the plate. Copper spray is also adequate and easier to apply in a thin coat.

With all the parts we have been throwing at this engine build, soon we will have a completed engine assembly that will look much like this engine running on the engine dyno. This particular stock short-block 5.7-liter LS1 with AFR 205 heads and a medium-sized 228/228 0.588-inch/0.588-inch 114 lsa +2 camshaft made 520 horsepower.

When you use a double-roller timing chain, it is necessary to space the oil pump 3 millimeters farther away from the engine to have enough clearance for the width of the timing chain. Coat the spacer with a thin coating of RTV or copper spray.

Install the spacer with the transfer hole aligned to the oil gallery holes in the block and the oil outlet hole in the oil pump. Slide it into place after loosely setting the oil pump into place.

The double roller chain, in addition to the oil pump spacers, will come with longer bolts and washers. With either factory or the longer bolts, dab some Loctite on the threads and install them, paying attention to center the pump to the oil pump gear, as in the photo nearby. The oil pump may be centered by using two 0.002-inch feeler gauges opposite each other between the oil pump housing, gears, and oil pump drive gear. Otherwise, the oil pump housing will slightly rub the oil pump drive. Install the spacers and bolts at the same time, making sure to line up the oil transfer hole completely with the oil pump outlet and block gallery hole, as it is offset compared to the oil pump mounting bolts. Once everything is in place and the oil pump is centered, you can first tighten up all the mounting bolts to seat the pump against the block. Then torque all four oil pump bolts to 18 lb-ft. Make sure the pump is secure to the block and that the oil pump bolts are not bottomed out, as some blocks have less threads available than others. You do not want a loose oil pump; it's somewhat obvious what damage that will cause.

Torque the oil pump bolts to 18 lb-ft while centering the oil pump on the crankshaft drive gear.

WINDAGE TRAY, SPACERS, AND CLEARANCE

A windage tray is an integral component that helps alleviate oil misting and minimizes oil splashing into the rotating assembly. The windage tray you use needs to match the oil pan and pickup tube it was designed for, as there are at least four variations of LS-series factory windage trays. With longer-than-stock stroke crankshafts or aftermarket connecting rods, clearance to the windage tray becomes a minor issue. It's better to spend the time carefully shimming the windage tray than to just leave it off; LS-series engines need all the help they can get to alleviate crankshaft windage. The least amount of oil impacting the crankshaft always results in more engine horsepower.

The windage tray installation is a straightforward on stock-stroke applications. Just bolt it on and go, normally. Any time the windage tray is installed, however, the rod-to-tray clearance should also be checked ritualistically. Ideal minimum clearance is 0.050 inch, although a 0.030-inch clearance is an absolute minimum if that's all that you can achieve. Use a flashlight to check each rod pin and corresponding outboard connecting-rod bolts, checking for interference concerns. Physically measuring the clearance under the tray is sometimes impossible. You can eyeball the clearance if required. Keep in mind that the clearance required is similar to the thickness of the average common padlock key as a visual reference.

Prepare the oil windage deflector tray for the larger diameter ARP main studs if necessary by enlarging the three smaller holes in the tray to 1/2-inch diameter. Using a tapered Unibit makes quick work of this procedure. Drilling these larger also helps you position the tray for rod bolt clearance. Deburr any sharp edges.

With longer-stroke crankshafts, something needs to be done to gain clearance for the windage tray. The easy thing to do is to stack two to three washers on each studded main bolt and add or subtract shims as needed.

Set the modified oil deflector tray into place and loosely tighten the included ARP tray nuts or OEM tray nuts if stock main bolts are used.

Torque the ARP or OEM deflector tray nuts to 18 lb-ft on all but the oil pickup tube bracket locations.

If you have adequate clearance to the windage tray, you can skip this section and move onto the oil pickup tube, but if clearance is tight, you need to compensate for it somehow. This is accomplished with normal 8-millimeter or 5/16-inch flat washers. You'd usually need to add two or three shims on each outer studded main cap bolt, adding or removing shims as necessary to gain proper clearance. Keep the same thickness of shims on the left or right row of studs.

Crankshafts with longer than 4-inch stroke throws may require that the windage tray mounting holes be opened up slightly to aid in positioning the tray in the best location for clearance. This is easily done with a common unibit to enlarge all the holes one size larger for better adjustment and positioning. You'll need to enlarge the mounting holes when using an ARP main stud kit, as the mounting bolts are 7/16-inch diameter versus the factory 8-millimeter (5/16-inch) setup.

If you are at three washers on each studded main cap bolt and need more clearance, mark these areas with a black marker and remove the windage tray. You will need to ding these areas, using a vice and the back side of a ball peen hammer to create additional room as required. You may find that after forcefully manipulating the windage tray with a hammer, you can remove some shims and keep proper clearances. The less shims the better. Once the windage tray is fitted properly, torque the windage tray nuts to 18 lb-ft and verify clearance one last time. Leave off the one to two windage tray nuts that support the oil pump pickup tube.

Check clearance between the tray and the connecting-rod bolts as the crankshaft is rotated. You will need a flashlight to see under the tray. Clearance between offending rod bolts and the tray should be at least 0.030 inch, but more is better.

OIL PICKUP TUBE, O-RING, AND TWEAKING

The oil pickup tube has the important task of transferring oil from the oil pan sump to the oil pump. The pickup tube also needs to match the windage tray and oil pan design. There are quite a few variances depending on oil pan and chassis setup.

On LS-series engines the pickup tube uses an O-ring seal at the pump. The O-ring seal dimension is dependent on the pickup tube's flange design. There are three styles of O-rings. Most F-body pickup tubes use the standard blue O-ring. If the oil pickup O-ring flange is straight with no O-ring recessed groove, the blue O-ring is required. If there is an O-ring recess on the flange, then normally the larger diameter green O-ring is required. There should not be a loose gap behind the O-ring, but it should also not be an extremely tight fit.

The oil pickup tube supporting bracket will also need to be modified for use with ARP main studs. Enlarge this hole using a Unibit to 1/2-inch diameter and make sure to deburr the sharp edges.

Place the oil pump O-ring on the pickup tube location and install the pickup tube into the oil pump. Shown here is a problem, though. I installed this O-ring dry to show how the O-ring will roll out of place as happens so commonly. Make sure the O-ring is fully seated in place; when building the engine on a stand there is no reason to mess this simple thing up.

The better way to install the oil pickup tube O-ring is to generously lubricate it and the inlet on the pump with assembly lube or anything slick.

The O-ring is a problem area and often low oil pressure concerns are a direct impact of O-ring installation problems or failure. Most O-ring failures are due to careless installation or reusing an old O-ring. When the pickup tube O-ring seal fails, it will usually result in low oil pressure and ticking valvetrain noises because the oil pump will be sucking up air bubbles along with the oil. The easiest way to install the O-ring and pickup tube assembly is to use a generous amount of assembly lube on both the O-ring and oil pump inlet. It should fall right into place with light pressure and have typical O-ring resistance as it is compressed. Line up the support bracket to the matching main stud and thread the main stud nut on to hold the oil pickup tube into place. Use Loctite on the pickup tube flange mounting bolt and torque it to 106 lb-in. Then torque the final windage tray and pickup tube bracket nut(s) to 18 lb-ft.

Using assembly lube, the O-ring and tube slide into place easily. Using this method will help ensure the O-ring is in its correct location and common oil pickup issues will be alleviated. Use a new O-ring any time this seal is broken.

Sometimes if you use spacers for the windage tray to clear the rotating assembly, the pickup tube screen is in turn relocated closer to the bottom of the oil pan. Mock up the oil pan without the oil pan gasket and make sure the pickup tube is not against the bottom of the oil pan sump. One-quarter inch to 3/8 inch of clearance is required to safely supply oil to the engine. You can use also use a small blob of clay on the pickup tube, install the pan, and measure how thick the compressed clay is. Usually light hand pressure on the oil screen itself will suffice for enough clearance normally. This is to offset the amount of shim under the pickup tube and windage tray mounting locations, due to the spacers moving the pickup tube closer to the bottom of the oil pan sump. Don't install the oil pan just yet though.

Since we spaced the oil deflector tray downward, it in turn affects the oil-pickup tube location. Place clay or a washer or a nut of known dimensions on the pickup tube screen to measure the clearance.

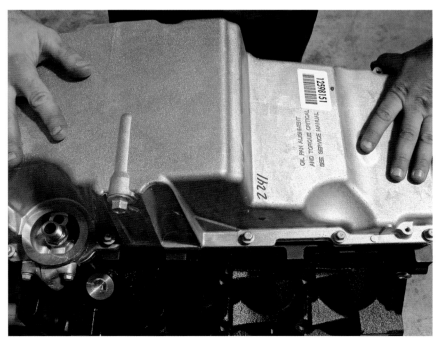

Set the pan into place. If the object suspends the oil pan above the block, the clearance is too tight and the pickup tube screen will need to be manipulated for clearance. The pickup tube clearance to pan needs to be at least 1/4 to 3/8 inch.

CHECKING F-BODY OIL PAN CLEARANCE NO. 1 ROD BOLT

With long-stroke engines, the oil pan sometimes needs modification. This is normally the case with F-body oil pans as the front section of the oil pan is shallow to clear the engine k-member and steering rack. F-bodies also use a four-fifths-length windage tray, while all other applications use a full-length windage tray. If you are using the full-length windage tray and the rod bolts clear it, it is unnecessary to check the oil pan clearance against rod bolts; however, the windage tray to oil pan clearance should be checked.

Now that you are familiar with the clearance issues of longer stroke crankshafts, install the oil pan with just four bolts toward the front securing the pan to the block. Slowly rotate the crankshaft, noting if there are any hard spots in the rotation. You can sneak a peak of the problem area if you use a small flashlight and watch the no. 1 connecting rod as it passes the oil pan while rotating. Four-inch stroke crankshafts are normally okay here, depending on the connecting-rod style. When you get to the 4.100-inch and larger stroke crankshafts, modifications are almost always necessary for adequate clearance.

It's hard to physically measure, but eyeballing the clearance on the no. 1 rod bolt should suffice. If it rubs the oil pan, or the rod bolt hits the oil pan and stops the crank rotation by hand, then more attention is definitely required. Either grinding or cutting a clearance groove is required when this happens. Worst case is that the oil pan will need a notch

When using F-body oil pans on strokers, the no. 1 connecting-rod bolt sometimes interferes with the oil pan itself. The oil pan will need to be ground down or notched to create the necessary clearance.

cut from it, and then re-welded further out to gain clearance. It's not unusual to remove material from any cast-aluminum oil pan to allow for proper clearances in this problematic area because of extra crankshaft stroke and extra girth on the big ends of the connecting rods.

Clearancing the pan by grinding is the easier thing to do if enough material is present. If the pan needs to be notched more, a hole can be cut here to be replaced by a patch plate welded and spaced down in a lower location.

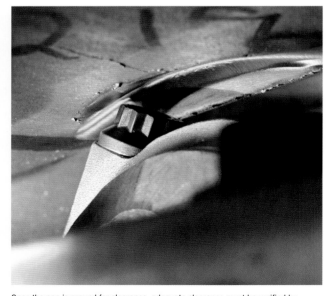

Once the pan is ground for clearance, adequate clearance must be verified by bolting the oil pan back into place and rotating the crankshaft while visually inspecting for adequate clearance with the timing cover left off. This might take a few attempts to gain proper clearance. Do not go crazy while grinding; just take out enough to clear. One could also use clay to measure for clearance if visual measurements are not trustworthy.

OIL PAN, TIMING COVER, REAR COVER INSTALLATION, AND ALIGNMENT

Gen III/IV LS-series engines all use a cast-aluminum oil pan and external engine covers that add additional structure and strength to the engine block case. This is the also the reason LS-series engines use captured O-ring, aluminum-housed engine gaskets in all locations. These solidify the engine's structure rather than insulating it, while also providing an almost 100 percent leak-proof engine assembly. If you have worked on a non-LS–powered vehicle, you will appreciate the lack of problematic oil leaks.

Installation of the oil pan is almost a no-brainer operation, although there are two things that require attention. First, install the oil pan gasket on the block if you're replacing it, using RTV sealer on the four gasket corners where it meets the edge of the block. Next, set the oil pan over the gasket, using care not to disrupt the oil pan gasket too much. The oil pan should sit flush on the oil pan gasket on all sides. If it doesn't, something is interfering with it. Inspect and correct that issue before tightening the oil pan bolts. Install all M8 oil pan bolts except the two long rear M6 bolts and the two front M8 bolts that extend into the timing cover. The rear

Another location where oil pan modification can be helpful is the oil filter inlet on the oil pan. Here a rather large casting bump restricts oil flow.

By removing the oil pan and opening this area up a tad, oil flow is improved. Some pans are worse than others. It is probably not that big of a restriction, but you do not want an oil restriction anywhere in your block if you can help it.

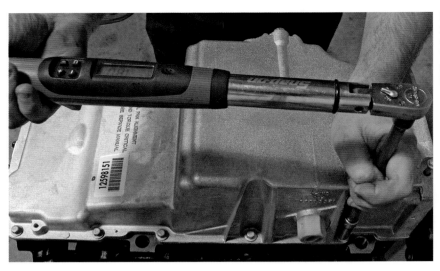

Finally, we can install the oil pan. The rear oil pan flange needs to be aligned to the bell housing flange of the block with a straightedge, and then the oil pan can be torqued. Torque the M8 oil pan bolts to 18 lb-ft in a circular pattern. If the rear cover is installed, you can torque the long M6 oil pan bolts to 106 lb-in.

flat flange of the oil pan now must be aligned to the block's bell housing flange. The pan rear surface needs to be within 0.010 inch front to rear of the block's rear surface. The easiest way to do this on the engine stand is to use a straightedge on each side of the engine to make sure both surfaces are equal. Once checked, you can now torque the M8 oil pan bolts to 18 lb-ft in a circular outward pattern, starting in the center. Finally, torque the last M8 bolt located rearward of the oil filter a little less. This bolt tends to crack the oil pan flange when fully torqued. Sixteen lb-ft are adequate.

If you have an F-body or C5/C6 bell housing removed from your transmission or torque tube, you can use that to additionally align the oil pan flange. You would have to remove the engine from the engine stand for this procedure. Once removed, install the transmission bell housing using four transmission bell housing bolts into the block and two bell housing bolts into the oil pan. Just tighten these by hand and then torque the oil pan bolts to specification. Some people install the oil pan last to use this oil pan alignment procedure. I prefer to install the oil pan using the straightedge method on the engine stand. I find it helps to have the oil pan installed to assist in the front and rear cover alignment, which is next.

After installing the oil pan, the front and rear covers can be installed, in either order. Depending on your engine stand, your rear cover may not fit between the stand and the engine. If so, wait until you remove the engine to install the rear cover. If sufficient room is available to slide the cover through the engine stand, then apply dabs of silicone to the corners of the rear cover gasket and install the rear cover.

With the oil pan installed, you can install the front cover easier as you now have a flat lower location with which to align the cover. All you need to worry about now is the alignment left to right. GM dealerships use a tool that centers the front and rear covers to the centerline of the crankshaft, which is a great tool to use.

Use RTV silicone where the front cover meets the block and oil pan. You can initially torque the eight timing cover bolts and two front oil pan bolts to 18 lb-ft.

I used spacers to leave enough room to install the rear cover while on the engine stand. This is not normally the case. If spacers were not used, one would have to remove the engine from the stand to install the rear cover.

Install the rear cover, allowing the rear main seal to center the rear cover. Like the front cover, it helps to use RTV silicone in the parting gap where the rear cover gasket meets the block and oil pan. When the rear cover is drawn into place, tighten the oil pan M6 bolts snug and then torque the rear cover bolts to 18 lb-ft. Then torque the M6 pan bolts to 106 lb-in.

The OEM front and rear covers need to be centered to the crankshaft centerline, as they are not doweled like other engine designs. Alignment can be done with the GM alignment tool, but using the rear seal itself has proved sufficient. There are two basic styles of rear main seals, which can be installed at different times. If you have the newest-design GM rear-facing lip seal, it can be installed into the rear cover before or after installing the cover, although if installed beforehand, it helps with rear cover lateral alignment. The forward-facing lip seal needs to be installed after installing the rear cover, using either the GM install tool or carefully guiding the sealing lip into place and tapping it into the cover with a small hammer and spacer, such as an old rear seal.

Start all 12 M8 rear cover bolts by hand and install the two long M6 oil pan bolts into the rear cover, keeping them tight enough for the cover to move around if needed. Alternate between tightening the lower two rear cover bolts and the two M6 oil pan bolts until the corner gap is null. Then tighten the rest of the bolts to seal against the rear cover gasket. Torque the rear cover bolts to 18 lb-ft and the two M6 oil pan bolts to 106 lb-in.

The Gen III valley plate is quite easy to install. Just lay the gasket on the block first, making sure its profile matches the block.

The timing cover may be installed next. Installation is similar to the rear cover. Install the timing cover gasket with RTV silicone on the corner ends and install the timing cover afterward. When using a double-chain and oil pump spacers, some oil pumps may physically interfere with the timing cover. If so, the timing cover will have to be ground down for clearance. Using clay or a marker to find the tight areas, it's a trial-and-error process of grinding, checking, grinding, and installation. Once completed, thoroughly clean the cover of aluminum debris and install the eight M8 timing cover bolts and the two M8 front oil pan bolts. Tighten the lower bolts and oil pan bolts first to pull the cover into place; verify the cover location and then tighten and torque the bolts completely.

There are a few different versions of the LS-series timing cover, and also aftermarket two-piece covers are now available. The Gen III OEM front covers are the same between all makes of vehicles, but the Gen IV covers are where things get a little complicated as the LS2, LS7, and L92 all use different covers, not to mention the front-wheel-drive Gen IV variants. If building a Gen III block, use the Gen III front cover. If building a Gen IV based block for a Gen III vehicle, use the LS2 front cover. With all other vehicles, it's probably best to use what originally came

These two grommets seal the valley plate to the knock sensor pedestals. They are commonly loose in used covers, and new grommets will need to be installed.

To install the valley plate, just align the knock sensor grommets and push down into place. Torque the valley plate bolts to 18 lb-ft. The Gen IV cover is similar, but make sure the proper displacement-on-demand (DOD) O-rings are in place to seal oil pressure from the DOD towers.

equipped with your Gen IV engine to avoid confusion and extra re-installation time if not correct.

The valley plate is the easiest cover to install. It is nowhere near any moving parts. Make sure you have the right cover. There are two Gen III covers: the LS1 standard cover and the LS6 with the valley plate PCV system. The PCV-improved LS6 cover can fit LS6-based 5.7-liter blocks and with light machining can be used on the original LS1 blocks. The LS6 cover will also fit the 6.0-liter Gen III truck block with no machining. For Gen IV engines, there are at least three main valley cover designs, depending on if the vehicle is a car, truck, or has active fuel management (or DOD). The standard LS2/LS3 or LS7 valley plate cover is needed on most Gen IV–based applications.

Keep in mind to double and triple check the Gen IV valley plate covers to make sure the eight DOD O-ring block-off seals are in place. Missing O-rings will significantly cause lower engine oil pressure throughout the engine via an internal engine oil hemorrhage. The only Gen IV blocks that do not need to use the O-rings are the LS7 block, and the GMPP LSX block as the DOD towers are not machined or existent in these blocks.

Installing the rocker covers is a straightforward procedure also. Make sure to pre-lube the valvetrain before installing the rocker cover.

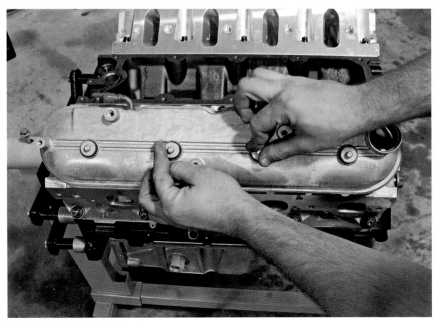

The only minor headache here would be if the aftermarket rocker arms physically interfered with the cover. The options at that point would be to use rocker cover spacers or grind the cover as required. If no interference issues are present, proceed to install the rocker cover bolts.

VALVE COVERS AND SPACERS FOR THE GEN III AND GEN IV

Installing rocker (valve) covers is usually a painless procedure, worth mentioning only for the variations in design. The 1997–1998 LS1 engines used perimeter bolt-mounted valve covers. If you are in the stone age and using these heads, the matching valve covers must be used as well. All center-bolt cathedral-port heads can use 1999 and newer valve cover designs. Rectangle-port heads with offset rocker arms require the use of a 2005 and newer passenger-side (right) valve cover because the Gen III valve cover baffles don't clear the offset intake rocker arms. Generally speaking, if you're replacing a Gen III with a Gen III head, you'll have no issues. Likewise, when replacing a Gen IV with a Gen IV head or Gen III head, you don't need to address any fitment issues. The only conflict is when you use a Gen IV rectangle port head with Gen III valve covers.

Most of the stock-replacement "bolt-on" aftermarket rocker arms such as Harland Sharp, Yella Terra, Competition Cams, or Scorpion, are typically designed for factory valve covers, but if clearance issues present themselves, you can either purchase aftermarket valve covers or valve cover spacers. With shaft-mount rockers, valve cover spacers or taller rocker covers are a requirement in all instances.

CAM AND CRANK SENSORS, KNOCK SENSORS, TRANS DOWELS, AND PILOT BEARINGS

To finalize installation preparation, the last things to install are the external sensors and transmission dowels, if you used a new block. The Gen III cam and crank sensors are a direct fit into the block itself. Just install and torque the M8 retaining bolts to 18 lb-ft. The Gen IV crankshaft sensor is similar, but the camshaft sensor is mounted in the timing cover instead of the block. The Gen III engine knock sensors mount in the two wells located in the center valley plate; these torque to 15 lb-ft. In a Gen IV, the knock sensors mount to the side of the engine block, one on each side. Torque these to 15 lb-ft. When retrofitting a Gen IV into a Gen III vehicle, the Gen III knock sensors need to be used. An open M10 bolt hole location can be used, such as on the side of the block.

Locations depend on application and type of headers, and they block access clearance; but a knock sensor harness extension must also be procured or built. The transmission dowels are simple and just need to be tapped in with a small hammer until they are in place.

With some vehicles it is best to wait until the engine is in place before installing the crankshaft pulley for maximum installation clearance. To install the pulley, you need either a GM installation tool or a makeshift threaded installation rod.

With the rocker cover installed, you can evenly torque all the rocker cover bolts to 106 lb-in. Do not go tighter than that setting; the small M6 bolts break easily.

The threads are M16x2.0. Locate a local bolt bin where you can purchase a long threaded rod that you can cut to length, a few M16x2.0 nuts, and a stack of large washers. Install the pulley over the crankshaft, making sure your new front seal is in place first. Then insert your installation tool and press the crankshaft pulley into place until it stops against the lower crankshaft gear. Remove the installer and, using an old crankshaft bolt, attempt to torque the bolt to the GM-specified 240 lb-ft to fully seat the pulley.

Yes, I know it says 240 lb-ft, but this is just about impossible for a normal human to achieve without an 8-foot breaker bar. Once you give up on that torque setting, and you're confident your pulley is installed completely, take the bolt back out. Install your new crankshaft bolt and torque it to a first pass setting of 37 lb-ft. I recommend using Loctite on the threads. With the crankshaft or flywheel locked down, torque the crankshaft pulley bolt an additional 140 degrees. The proper way to lock down the engine to install the pulley bolt is to have the flywheel in place and use the GM flywheel holding tool from a GM dealership. Most people won't have access to that, so the alternative is to install two opposing ARP flywheel bolts and use a long pry bar against the engine stand or bell housing bolt instead.

One of the last items to install is the crankshaft damper pulley assembly. This requires the use of a special threaded installer. Make sure to align the keyway to the pulley if a keyway is present; otherwise, it is just a press fit, like the OEM pulley.

The use of an M16x2.0 threaded rod, washers, and a M16 nut is required to correctly press the crankshaft damper into place. If you use a long bolt instead, you run the risk of permanently stripping or damaging the crankshaft threads. Spending $20 to build your own installer is an inexpensive alternative.

At this point, your engine is finally ready for installation. Of course, things such as exhaust, intake, and remaining vehicle-specific driveline components must be installed as well. With any LS-series engine, the installation does not vary much from the OEM engine. As long as the OEM engine is of the Gen III or Gen IV family, it's a bolt-in deal.

When the engine is installed, use a good break-in oil, or use conventional non-synthetic engine oil and add your own break-in additives. Break-in is crucial to an engine having a good, long life. Joe Gibbs Racing offers racing bred oil formulas made for break-in and racing purposes. Modern "shelf" oil mixtures are required by the Environmental Protection Agency to have less zinc and phosphates to prevent catalytic converter failures, but these additives are needed greatly for a fresh engine to prevent scuffing on the pistons and rings. Clearances are tight and friction is greatest on a brand-new engine during engine break-in.

When you're ready to fire up the new engine, it's good practice to leave the fuel pump fuse out to bump the starter over a few times just to verify no mechanical interference problems slipped by. Once you rotate the engine a few times, you can replace the fuel pump fuse and attempt to fire the engine up. Remember that if you changed fuel injectors to match a larger engine displacement, the PCM tuning will need to be changed to reflect the injector sizing. If not, too much fuel can wash the engine oil off the cylinder walls, and accelerated piston ring wear can occur quickly.

With the crankcase full of your required engine oil, the oil filter pre-filled with oil, and the cooling system full of coolant, you can start up the engine. Let it build oil pressure on its own, and once you've verified that oil pressure rises quickly, look under the car for oil or coolant leaks. If any problems arise, shut the engine off right away and repair the concern. If no problems are found, hold the engine rpm in the 1,500 to 2,000 range to help sling oil onto the cylinder walls and camshaft to prevent scuffing. Hold the engine's rpm in that range until it warms up to temperature for 5 to 10 minutes. Shut off the car if the oil pressure is too low or coolant temperature is too high; keep an alert eye to your gauges at all times.

When the engine is fully warmed up, shut it off and let it cool down. Then change the oil once the engine has cooled down to comfortable levels.

It is crucial anytime a new setup is installed to have the PCM tuning gone through to make sure everything is right. Some builders prefer at least 500 miles on the engine before running it hard, while others have had good results just romping on it from the get-go after initially changing the engine oil and filter. There is really no wrong way to go about this, but it is more of a personal choice in the matter whether to wait or not.

The OEM crankshaft center bolt uses a strenuous torque procedure. First, an old bolt is torqued to 240 lb-ft to seat the balancer and lower timing gear. Then you remove the old bolt and install the new, untorqued bolt. I prefer to use red Loctite on this bolt as additional insurance.

Torque the new crankshaft bolt to 37 lb-ft, and then angle-torque the bolt to 140 degrees to ensure it does not budge.

CONCLUSION

The Gen III and Gen IV small-block engines may have had an insignificant start before anyone had their hands on them, but once word traveled on the street, racetracks, and the key Internet forums (such as www.LS1tech.com). Web-based forums helped coax along the drive for more performance and provided a hangout where F-body fanatics could discuss their newfound power modifications and compare most recent time slips with LS-series peers. This sometimes friendly competition is the stamina that has helped keep this engine family thriving for modifications for more than 10 years. The LS-based engines have breathed new life into late-model performance, and they are destined to become the engines that you'll tell your grandkids about.

Not a day goes by where you don't hear someone say something to the effect of "I can't believe how much power those LS1s are making now." It is true. These are one efficient overall package that overperform and never let down. Just looking at quarter-mile performance, there are seven-, eight-, and nine-second F-bodies becoming more and more commonplace. There is a six-second pass held by Kurt Urban Performance (formerly of Wheel to Wheel) using a turbocharged Gen III–based engine setup. Several stock 5.7-liter short blocks in both F-body and C5 running into the nine-second range, while staying naturally aspirated.

While most of what you may see entails Camaros and Trans-Ams due to popularity and ease of accessibility into these vehicles, there are also breeds of performance-hungry Corvette owners of both C5 and C6 'Vettes that also have almost the same power-drive as their F-body cousins. With GTOs, Silverados, CTS-Vs, Monte Carlos, Impalas, G8s, and the 2010 Camaro (on the horizon), you can be assured that the GM LS-series engine and its future offspring spinoffs will be in GM performance vehicles for years to come.

LS-series Engine - Torque and Bolt Specifications: 1997–2009 Gen III / Gen IV Engines — Joseph Potak

Type and Location	Diameter	Thread	Bolt Length	Fastener Size	Torque Spec.	Quan.	Engine	Misc.
OEM Camshaft Plate Bolts	8mm	1.25	20mm	10mm	18 lb-ft	4	All	Using Loctite
ARP Camshaft Plate Bolts	8mm	1.25	20mm	10mm	18 lb-ft	4	All	Using Loctite
Torx Camshaft Plate Bolts	8mm	1.25	18mm	T-40 Torx	11 lb-ft	4	All	Countersunk Bolts
GMPP LSX Camshaft Plate Bolts	8mm	1.25	20mm	5mm Allen	15 lb-ft	6	LSX Only	Button Head Allen
OEM Camshaft Gear Bolts	8mm	1.25	24mm	10mm	24 lb-ft	3	All	Using Loctite
ARP Camshaft Gear Bolts	8mm	1.25	25mm	10mm	28 lb-ft	3	All	Using Loctite
OEM Lifter Tray Bolt	6mm	1.00	19mm	10mm	106 lb-inch	4	All	Shouldered Bolt
GMPP LSX Lifter Tray Bolt	6mm	1.00	24mm	10mm	106 lb-inch	8	LSX Only	Exclusive LSX Bolt
OEM Rocker Arm Bolts	8mm	1.25	45mm	8mm	26 lb-ft	16	All	Use RTV for intake threads
Allen Rocker Arm Bolts	8mm	1.25	45mm	6mm Allen	24 lb-ft	16	All	Use RTV for intake threads
LS7 Rocker Arm Bolts	8mm	1.25	35mm	8mm	24 lb-ft	16	LS7	
Center-bolt Rocker Covers	6mm	1.00	Unique VC bolt	8mm	106 lb-inch	8	All 1999+	
Oil Pump Mounting	8mm	1.25	30mm	10mm	18 lb-ft	4	All	Rec. Loctite
Oil Pump Plate	6mm	1.00	13mm	10mm	106 lb-inch	7	All	
Oil Pickup to Pump	6mm	1.00	16mm	10mm	106 lb-inch	1	All	Rec. Loctite
OEM Windage Tray Nuts	8mm	1.25	Nut	13mm	18 lb-ft	8-10	All	
Oil Pan Flange Bolts	8mm	1.25	30mm	10mm	18 lb-ft	12	All	
Long Oil Pan Bolts	6mm	1.00	136mm	10mm	106 lb-inch	2	All	
Front Cover Bolts	8mm	1.25	30mm	10mm	18 lb-ft	8	All	
1st Design Rear Cover Bolts	8mm	1.25	30mm	10mm	18 lb-ft	12	All	

Type and Location	Diameter	Thread	Bolt Length	Fastener Size	Torque Spec.	Quan.	Engine	Misc.
2nd Design Rear Cover Bolts	8mm	1.25	25mm	10mm	22 lb-ft	12	All	
Gen III / Gen IV Valley Cover	8mm	1.25	30mm	10mm/13mm	18 lb-ft	10-11	All	
Water Pump Bolts	8mm	1.25	92mm	10mm	18 lb-ft	6	All	
Timing Chain Damper	8mm	1.25	38mm	13mm	18 lb-ft	2	Mainly Gen IV	
LSX Crank Damper Bolt	16mm	2.00	103mm	24mm	1st Pass 37 lb-ft 2nd Pass 140°	1	Except LS7	After balancer seated Use NEW bolt w/Loctite
LS7 Crank Damper Bolt	16mm	2.00	126mm	24mm	1st Pass 37 lb-ft 2nd Pass 140°	1	LS7	After balancer seated Use NEW bolt w/Loctite
Early OEM Rod Bolts	9mm	1.00	43mm	11mm	1st Pass 15 lb-ft 2nd Pass 60°	16	Pre-2000	Using Oil in pairs
Late OEM Rod Bolts	9mm	1.00	43mm	11mm	1st Pass 15 lb-ft 2nd Pass 75°	16	Post-2000	Using Oil in pairs
LS7 OEM Rod Bolts	9mm	1.00	35mm	12-point 10mm	1st Pass 15 lb-ft 2nd Pass 110°	16	'06+ Titanium Rods	Using Moly Lube in pairs
Aftermarket Rod Capscrews	7/16"	Fine	1.600"	12-pt 3/8" or 7/16"	Refer to Manufacturer or ARP	16	Any	Using ARP Lube/Grease
OEM Inner Main Cap Bolts	10mm	2.00	102mm	13mm	1st Pass 15 lb-ft 2nd Pass 80°	10	All	Using Oil in sequence
OEM Outer Main Cap Bolts	10mm	2.00	94mm	15mm	1st Pass 15 lb-ft 2nd Pass 53°	10	All	Using Oil in sequence
OEM Side Main Cap Bolts	8mm	1.25	20mm	10mm	18 lb-ft	10	All	Use RTV on bolt head in sequence
ARP Inner Main Cap Stud Kit	10mm	1.25	Nut	12-pt 12/17mm	60 lb-ft	10	All	Using ARP Lube
ARP Outer Main Cap Stud Kit	10mm	1.25	Nut	12-pt 12/17mm	50 lb-ft	10	All	Using ARP Lube
ARP Side Main Cap Bolts	8mm	1.25	20mm	12-pt 10mm	19 lb-ft	10	All	Use RTV on bolt head
ARP Windage Tray Nuts	10mm	1.25	Nut	14mm	18 lb-ft	8-10	All	
OEM Cylinder Head Bolts 1st Design Longest Length	11mm	2.00	156mm	15mm	1st Pass 22 lb-ft 2nd Pass 90° 3rd Pass 90°	16	Pre-2004	1st design position 1-8 in sequence
OEM Cylinder Head Bolts 1st Design Medium Length	11mm	2.00	100mm	15mm	1st Pass 22 lb-ft 2nd Pass 90° 3rd Pass 50°	4	Pre-2004	1st design position 9-10 in sequence
OEM Cylinder Head Bolts 2nd Design Medium Length	11mm	2.00	100mm	15mm	1st Pass 22 lb-ft 2nd Pass 90° 3rd Pass 70°	20	Post-2004	2nd design in sequence
OEM Cylinder Head Bolts	8mm	1.25	45mm	10mm	Final Pass 22 lb-ft	10	All	Intake Flange Row
ARP Head Bolt Kit	11mm	2.00	156mm/100mm	1/2" 6-pt	1st Pass 30 lb-ft 2nd Pass 50 lb-ft 3rd Pass 70 lb-ft	20	All	Using ARP Moly Lube in sequence
ARP Head Stud Kit	11mm 7/16"	2.00 20NF	Stud Nut	12-pt 1/2" or; 6-point 5/8"	1st Pass 30 lb-ft 2nd Pass 50 lb-ft 3rd Pass 80 lb-ft	20	All	Using ARP Moly Lube in sequence
ARP Intake Row Flange Bolts	8mm	1.25	Stud/Bolt	12-pt 3/8"/10mm	Final Pass 23 lb-ft	10	All	Using ARP Moly Lube
LSX Block ARP 2000 Studs *GMPP LSX Specific	11mm 7/16"	2.00 20NF	Stud Nut	12-pt 1/2"	1st Pass 30 lb-ft 2nd Pass 60 lb-ft 3rd Pass 105 lb-ft	20	LSX 6-bolt Heads	Using ARP Moly Lube in sequence
LSX Intake Row Flange Bolts	8mm	1.25	Stud	12-point 3/8"	Final Pass 28 lb-ft	10	LSX	Using ARP Moly Lube
LSX Block 5th/6th Head Studs	8mm	1.25	Nut	3/8"	Final Pass 28 lb-ft	32	LSX Block Only	Using ARP Moly Lube

Index